1 9 96

Agricultural and Environmental Research in Small Countries

Agricultural and Environmental Research in Small Countries

INNOVATIVE APPROACHES TO STRATEGIC PLANNING

PABLO EYZAGUIRRE
International Plant Genetic Resources Institute (IPGRI), Italy

JOHN WILEY & SONS
Chichester • NewYork • Brisbane • Toronto • Singapore

1996

Other Wiley Editorial Offices

John Wiley & Sons, Inc., 605 Third Avenue,
New York, NY 10158-0012, USA

Jacaranda Wiley Ltd, 33 Park Road, Milton,
Queensland 4064, Australia

John Wiley & Sons (Canada) Ltd, 22 Worcester Road,
Rexdale, Ontario M9W 1L1, Canada

John Wiley & Sons (Asia) Pte Ltd, 2 Clementi Loop #02-01,
Jin Xing Distripark, Singapore 0512

Library of Congress Cataloging-in-Publication Data

Eyzaguirre, Pablo B.
    Agricultural and environmental research in small countries :
  innovative approaches to strategic planning / Pablo Eyzaguirre.
        p. cm.
    Includes bibliographical references and index.
    ISBN 0-471-96074-8 (ppc)
    1. National agricultural research systems—Developing countries—
  Planning.   2. Agriculture—Research—Developing countries—
  Planning.   I. Title.
  S542.3.E975   1996   96-6208
  630'.720172'3—dc20   CIP

British Library Cataloguing in Publication Data

A catalogue record for this book is available from the British Library

ISBN 0-471-96074-8

Produced from camera-ready-copy supplied by author
Printed and bound in Great Britain by Biddles Ltd, Guildford and King's Lynn
This book is printed on acid-free paper responsibly manufactured from sustainable forestation,
for which at least two trees are planted for each one used for paper production.

Contents

Preface

The basic definition of small developing countries in ISNAR's long-term study is linked to a population of fewer than 5 million people, an agricultural economy, and a per capita income of less than US$ 2000. Most countries falling into this classification face the same (and, often, more) problems related to agricultural development and agricultural research that challenge larger countries. Many of the key functions of technology generation and adaptation that are central to national systems in larger countries are difficult or impossible to undertake or manage in small ones. Schumacher's contention that small is beautiful can hardly be applied to the development problems of small countries.

On the positive side, at the outset of this study, it was felt that some countries had overcome at least some of the constraints associated with small size and that much could be learned by studying their success.

Farmers and their service organizations, agricultural scientists, and policymakers in small countries are fully aware of their large problems. Furthermore, because of their size, they face unique difficulties in dealing with technical agencies, financial donors, and other partners. In general, their research systems are weak, their potential contributions to large problems are not great, transaction costs are high, and the chances of sustaining their national institutions are limited.

In 1989, ISNAR started a long-term collaborative study when it was recognized that in addition to the problems commonly faced by all developing countries, small countries face many challenges specific to their size. This study was conducted under the leadership of Dr. Pablo Eyzaguirre and included professionals from both small and large countries, from international agricultural research institutes, and many others who participated together with ISNAR's staff. A number of donors, especially the Government of Italy, provided financial support. ISNAR gratefully acknowledges all those who participated in, contributed to, and supported the "small-countries study."

A number of the study's different results and findings are already recorded in other publications. This book provides an overview and summary of the different aspects of the study as they relate to agricultural research in small countries. Tools and methods are highlighted to better identify the scale and to set the scope of agricultural research in these countries and to define relevant institutional requirements and possibilities. It is expected that not only will researchers and academicians benefit from the lessons we learned in this study, but that policymakers and research leaders in all developing countries will find ideas and conclusions that might be applicable in their own specific situations.

Christian Bonte-Friedheim
Director General
International Service for National Agricultural Research
December 1995
The Hague

Acknowledgments

This book contains and summarizes the ideas and work of a team of researchers and research leaders dedicated to building indigenous research capacity in small developing countries. Their initiative, experience, and insight provide the basis for the strategic approach outlined in this book.

National Case Study Consultants:
Ayele Gninofou and M. A. Tonyawo Aithnard (Togo)
Trower Namane (Lesotho)
Mohammed Dahniya (Sierra Leone)
Janice Reid (Jamaica)
Param Sivan (Fiji)
Mario Contreras (Honduras)
Jagadish Manrakhan (Mauritius)

Regional and Thematic Consultants:
Samsundar Parasram (Caribbean)
Elon Gilbert (West Africa)
Jaap Arntzen (Natural Resources)
Carlos Zacarias (Marketing)
Miguel Rojas (Diversification)

The ISNAR/University of Mauritius International Workshop on Management Strategies and Policies for Agricultural Research in Small Countries, held in May of 1992, brought together a host of research leaders and policymakers whose contributions (cited in the text) were crucial to the synthesis contained in this book. I thank the University of Mauritius and its Food and Agriculture Research Council for having helped to bring us together and crystallize some new strategic approaches.

External advisors from a number of international and technical agencies contributed invaluable support and wisdom at various stages in the project. I wish to give special thanks to Peter Matlon, Bishnodaut Persaud, Gora Beye, and Mohan Narain who provided guidance and support at crucial moments in this complex endeavor. A group of advisors within ISNAR contributed much experience and insight to ensure that the results of the study contributed to ISNAR's mission of strengthening national agricultural research systems. I am particularly indebted to three ISNAR colleagues who served as advisors to this project through all its phases: T. Ajibola Taylor, Carlos Valverde, and N'Guetta Bosso. At different stages and in different areas we received outstanding intellectual input and crucial advice from Dely Gapasin, Luka Abe, S. Huntington Hobbs, Gabrielle Persley, and Peter Goldsworthy.

ISNAR's leadership took a chance in supporting a study with a very broad scope that straddled fences and trampled disciplinary boundaries in search of new models and strategies for agricultural research. It also risked going counter to the established practice of collaboration and advice to national research organizations. For this I thank H. K. Jain for arguing that small-country research systems were indeed different and for thus helping to get this project underway. Howard Elliott's extraordinary intellectual energy, breadth, and acuity provided a source of support without which the project could never have been completed. Christian Bonte-Friedheim's profound commitment to serving those countries where the needs are the greatest afforded us the privilege and benefit of his leadership and participation in the key workshops of this project in 1990 and 1992. In him, small developing countries have a friend who continues to champion innovation and the need for more support to their research systems.

The ISNAR project team worked over three years to synthesize the results. This team included Peter Ballantyne, who directed the companion study on information systems and strategies and whose contributions are evident in every aspect of this book. Andrew Okello carried out the compilation and analysis of data on national research systems, supported case studies in the field, and co-authored several papers within this series. Lynette Thomas McGraith organized the documentation from the case studies and the final conference. Christine Solinger set up the procedures and structure to cope with a vast amount of information in running the project. Bob Solinger designed and taught us the use of various databases. Bonnie Folger McClafferty and Ornella Arimondo assembled the initial research data that defined the key countries and issues the project was to cover. We also had the benefit of working with Kathleen Sheridan as editor throughout the life of the project. She obliged us to ask questions and provide answers stated in clear and logical ways that people could understand. As part of her editorial team, Richard Claase gave inspired artistic and graphic support. Fionnaula Hawes' experience in production and layout was an added plus.

Completing a big and complicated study and producing this synthesis seemed daunting at various stages for our team. What made it possible and often enjoyable was the privilege of working with the fine people who were in various ways and at different times associated with the project. To all of them I give my sincere thanks.

                                                    Pablo Eyzaguirre
                                                    January 1996
                                                    Rome

Foreword

Five years after ISNAR was established to assist developing countries in strengthening their agricultural research systems, it became clear that small countries might need a different approach (de la Rive Box 1985). Their public-sector research organizations were often unable to muster sufficient financial resources or numbers of researchers to address the priority problems of the agricultural sector. The global agricultural research system that supports national research was either bypassing small countries or else overloading their research organizations with trials and projects that threatened to distort their national priorities.

Many of the key functions of technology generation and adaptation that were central to national systems in larger countries were proving difficult to undertake or manage in small ones. On the positive side, it was felt that some countries had undoubtedly overcome at least some of the constraints associated with small size, and that much could be learned by studying their successes.

These perceptions led ISNAR to launch a special project on the organization and management of agricultural research in small countries. This project, which began in 1990, was funded largely by the Government of Italy, but additional support came from the Rockefeller Foundation, the Danish International Development Agency, the Technical Centre for Agricultural and Rural Cooperation (CTA), the United Nations Development Programme (UNDP), and the British Development Division in the Caribbean.

The project focused on small but low- and middle-income developing countries where resource limitations and the inherent constraint of size restrict the scale of the research effort. It had the following objectives:

- to devise a means of measuring and classifying the key factors related to agricultural research in small countries so that the national research systems in this category could be analyzed and compared;
- to identify suitable research and organizational strategies for national research systems in small countries;
- to evaluate national and regional research environments in order to help small countries identify and exploit opportunities for acquiring new technologies from external sources;
- to seek ways in which national systems could improve their links with policy-makers, local producers, and external sources of knowledge and technology;

- to identify the skills that could enable research leaders in small countries to manage their research systems and institutions efficiently and effectively.

To implement the project, three types of studies were carried out. First, 50 small developing countries were selected for inclusion in a global database on the scale and scope of national agricultural research. These countries have populations of less than five million (1980 census) and meet at least three of the following four criteria:

- The economically active agricultural population was 20% or more of the total economically active population.
- Per capita income was less than US$ 2000 (1980 constant dollars).
- Agricultural gross domestic product (GDP) per capita for the economically active agricultural population was less than US$ 2000.
- Agricultural GDP was at least 20% of GDP.

Some developing countries have research systems that are small because they are at an early phase of their institutional development. These countries, many of which have explicit long-term plans to build larger research systems, were excluded from the study. For each of the 50 countries selected for inclusion, information on the structure, organization, and resources pertaining to research was used to assess the existing national research capacity in relation to the scope of research needs and activities.

Second, seven countries were selected for in-depth case studies: Honduras, Jamaica, Sierra Leone, Togo, Lesotho, Mauritius, and Fiji. The studies covered such topics as institutional development, research organization and structure, and external links and information flows.

Third, regional studies were conducted for the small countries of West Africa, the Caribbean, Southern Africa, and the South Pacific. The goal of these regional studies was to assess research capacity in regions where small countries predominate. The regional studies reviewed the participation of small national research systems in regional research and investigated whether they were gaining access to the information and technology they needed. The studies also considered the division of labor among national systems in a regional context, as well as the role of regional research organizations and collaborative networks.

In collaboration with CTA (EEC/ACP Lomé Convention) and agricultural research information specialists from developing countries, a companion study explored the management of scientific information in four small-country research systems: Swaziland, Trinidad and Tobago, Seychelles, and Mauritius. The rationale for this study was the importance of information management as a core function of a small-country research system. Information can indeed be used to supplement or replace some kinds of research, releasing scarce resources for programs that must be conducted locally. Furthermore, the effective scope of research in a country can be broadened by accessing and using relevant information.

Workshops have been an important tool for organizing this study and disseminating its results. The first workshop, held in The Hague in January 1990, reviewed project methodology and launched the country and regional studies. A regional workshop on "Strategic Planning for Small-Country National Research Systems" was also held in 1990 in Jamaica. This was for research leaders and regional organizations in the Caribbean. A similar workshop was held in 1991 in Western Samoa for the island nations of the South Pacific. A final workshop was held in Mauritius in May 1992, bringing together research leaders, information managers, and policymakers from 33 small countries. The papers and discussions presented and held at the Mauritius meeting provided much of the material for this book, but they remain unpublished; however, the case studies, issue papers, and conference proceedings produced in the course of the project have been published and are available from ISNAR.

Pablo Eyzaguirre
December 1995
Rome

1 Defining the "Small-Country" Problem

The constraint that small size places on the institutional and economic development of a country has been widely discussed. Studies of small developing countries have examined this issue in depth, showing that small can indeed be beautiful, but it is often fraught with danger and vulnerability (Schahczenski 1990). Many international organizations and development agencies have recognized that size is a constraint to the institutional and economic development of a country. In this book, we look at the effect of small size on the institutional structure and development of a country's national agricultural research system (NARS). Following this, we discuss the policies and organizational strategies that can be used to overcome size constraints.

Small is a relative term that can be applied to a wide range of countries, depending on how it is defined. One can use geographic size, population, or economic criteria. For development purposes, population and economic size are used to define a country's size. As Forsyth (1990) notes in his UNESCO study on technology policy:

> The most obvious and frequently used measure of size is population; countries differing in terms of population will, evidently, have different market sizes, different scale profiles of indigenous industry, different scope for labor specialization, different aggregate levels of savings and investment and so on. All of these may be expected to result in different performance patterns and different configurations and levels of technological development.

There is a growing consensus in international agencies and among development economists and planners that a population of 5 million is the limit below which the economy and institutions of a country are severely constrained and where some national institutions, infrastructures, and services become uneconomic (Commonwealth Secretariat 1990; Forsyth 1990; Jalan 1982). It is not surprising, therefore, that many of the countries considered to be "least developed" are also small countries.[1]

Small countries are diverse in many ways. They can be territorially vast like Mauritania, Chad, Botswana, Mongolia, Paraguay, or Namibia. They include island microstates like the Seychelles, Saint Lucia, and the Solomons, or small continental states surrounded by large neighbors such as the Gambia or Guyana. Some, such as Lesotho, Bhutan, and Swaziland, are landlocked, creating a special vulnerability in terms of trade and other links to the wider world.

They can be numerous within a region, as in Central America, the Caribbean, West Africa, or the South Pacific, or else widely scattered. Some, particularly the island nations in the Pacific and Indian Ocean, are distant from their markets and suffer problems of isolation.[2] While new information and transport technologies have gone a long way to reducing the impact of isolation on small countries, it nonetheless remains a significant feature of many small island nations.

For this study (see Box 1.1), we identified 50 low- and middle-income developing countries with small populations (fewer than 5 million in 1980) and significant agricultural sectors (see Figure 1.1), as indicated by contributions to GDP and the size of the agricultural labor force (Eyzaguirre 1991). These criteria exclude small but high-income countries like Brunei or Singapore, or those countries that are not significantly dependent upon agriculture.[3] Increasing agricultural productivity is crucial to the development of these small countries. To this end, they must build and manage viable agricultural research institutions.

Box 1.1. ISNAR Study on Small Countries

The ISNAR study on small countries began in 1990. It focused on agricultural research systems in small, low-, and middle-income developing countries. Its goal was to identify the strategies that research systems in small countries could adopt, and to understand the types and range of functions that they could perform.

A key project activity was to create a global data base on the scale and scope of agricultural research of NARS. For each country, information on the structure, organization, and resources pertaining to research was used to assess the existing national research capacity in relation to the scope of research needs and activities. The data base provides cross-country indicators of the organization and structure of NARS in small countries.

Global data were complemented by a series of national case studies in Honduras, Jamaica, Sierra Leone, Togo, Lesotho, Mauritius, and Fiji. The studies cover institutional development, research organization and structure, external linkages, and information flows to the country. Since small countries were expected to rely substantially on outside sources of technology, a series of regional studies covering West Africa, the Caribbean, Southern Africa, and the South Pacific were also carried out.

The critical role of information flows and their management was recognized in a companion study that explored the management of scientific information in small research systems.

IMPLICATIONS OF SIZE CONSTRAINTS FOR NATIONAL RESEARCH INSTITUTIONS

Problems associated with small size manifest themselves in four major areas. The first is the limit on the number and size of development institutions that small countries can sustain. The second concerns market isolation and the relative inability of a small-country economy to influence markets. The third area concerns the fragile

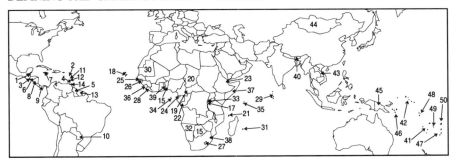

Figure 1.1. Small countries (as defined by this project):

Latin America and Caribbean	Africa and the Indian Ocean		Asia and the Pacific
1. Belize	15. Benin	28. Liberia	40. Bhutan
2. Dominica	16. Botswana	29. Maldives	41. Fiji
3. El Salvador	17. Burundi	30. Mauritania	42. Kiribati
4. Grenada	18. Cape Verde	31. Mauritius	43. Laos
5. Guyana	19. Central African	32. Namibia	44. Mongolia
6. Honduras	Republic	33. Rwanda	45. Papua New Guinea
7. Jamaica	20. Chad	34. São Tomé e	46. Solomon Islands
8. Nicaragua	21. Comoros	Principe	47. Tonga
9. Panama	22. Congo	35. Seychelles	48. Tuvalu
10. Paraguay	23. Djibouti	36. Sierra Leone	49. Vanuatu
11. St. Lucia	24. Equatorial Guinea	37. Somalia	50. Western Samoa
12. St. Vincent	25. Gambia	38. Swaziland	
13. Suriname	26. Guinea-Bissau	39. Togo	
14. Trinidad and Tobago	27. Lesotho		

or highly restricted natural resource base. The fourth is the vulnerability of a small country's economy and institutions (Commonwealth Secretariat 1990); their dependence on donors and external agencies is heightened and the relative impact of external factors is often greater.

DISECONOMIES OF INSTITUTIONAL DEVELOPMENT

Small countries tend to have smaller government-funded institutions of all kinds, both in agriculture and in other fields that are important for national development, such as education and health. Research institutes, government ministries, universities, and hospitals, along with the infrastructure, personnel, and equipment needed to make these institutions operational, are costlier to build and maintain in small countries than in countries that can spread these fixed costs over a larger population and economy.

Examples from public agricultural research institutions indicate that research investment per hectare or value of a commodity tends to be higher in small countries (Ruttan 1989: 340), and the case for greater investment in agricultural research is often hard to argue. These high costs are exacerbated by the agroecological diversity that typifies many of our sample countries.With few institutions, there are many crucial functions, including the training of scientists and other professionals, provision of specialized medical care, and even defense, for which small countries rely on external institutions or other countries.

The scale constraint on institutional development is inherent to small size. It is something that small countries have to accept as "in the nature of things". Building more or larger institutions to cater to development needs is not normally an option, unless windfall profits arise out of, say, the development of mineral resources. (Even then, institutional development and organizational growth is risky, since it may be difficult to sustain the new institutions after the "boom".) Instead, small countries must use new approaches to mobilize existing institutional resources in more flexible and diverse ways.

SMALL PRODUCTION AND MARKETS

Countries with small populations and low incomes have small internal markets, which means that many industries are uneconomic unless they can rely upon exports. These countries also have to import many goods and agricultural products that cannot be economically produced domestically. To pay for these imports, they are dependent upon export markets over which they have little influence.

The structure of the economies in many small countries is biased towards one or two export crops. A positive trade balance often requires some kind of externally supported special trade status such as that enjoyed by Barbados, Fiji, and Mauritius with sugar, and the favored status that Eastern Caribbean banana producers have received in the United Kingdom. This makes them vulnerable to agreements that can be changed, leaving them unprotected in the world economy. For some small states dependent upon a single export crop, the volume of their exports is so small in global terms that they have a difficult time marketing what they produce. This is the case of São Tomé e Principe, which relies on cocoa for over 90% of its export earnings, yet its annual production is less than 5000 tons. Even when new market niches can be found for high-value crops, a small country is not at a comparative advantage with respect to larger countries, which may have better transport and marketing structures (Poon 1990).

FRAGILE OR RESTRICTED RESOURCE BASE

Many of the small countries selected for our sample share the constraint of a fragile or limited resource base. Countries with a small population and large territory such as Namibia, Mauritania, Paraguay, or Mongolia have vast expanses of fragile lands that cannot support intensive settlement or use. Countries with a small territory and high population density can do little to increase the amount of land under production;

they now confront the classic Malthusian problems of overuse and degradation of their land, water, and plant resources due to population pressure and poverty, as is the case in Rwanda, Burundi, El Salvador, and Bhutan.[4]

Small countries with more intensive commercial or export agriculture face problems of pollution and resource degradation. For example, Mauritius, Swaziland, and Fiji are experiencing major problems in managing the wastes from their sugar and pulp industries, and small countries in Central America have to cope with the negative effects of pesticides in their fruit and vegetable export industries. Many small island nations depend on their marine resources, which are often undervalued or overexploited. Resource conservation is an important issue for all countries, but the economic effects of the inefficient use of environmental resources are felt sooner in small countries.

Most small countries do not yet have institutions that are able to address the complex scientific questions associated with these environmental issues, nor, in most cases, are they able to manage and direct the wealth of information needed to assess problems and reach decisions about their solutions. Much research on the management of natural resources is directed towards policy-making rather than towards the development of new technologies. The key factor here is how research feeds into the decision-making structures within government administrations. In this respect, a small country, where access to key decision makers may be less formally controlled than in large countries, may not be at a disadvantage.

Many natural resource management problems cross national frontiers. Besides the limited domestic capacity for research that small countries have in these areas, this is another reason why they will have to team up with larger regional and global research efforts in order to tackle these problems effectively.

VULNERABILITY OF INSTITUTIONS

We now have considerable experience showing that small countries and their institutions are inherently more vulnerable to instability or collapse (Commonwealth Secretariat 1990). This instability is nowhere more evident than in the political sphere. No national research manager in a small country needs to be reminded that his or her national research system operates in a climate in which the winds of political change can blow swiftly and, at times, highly destructively.

The margin for error in a small country or system is much smaller than in a large one. Small countries are more strongly affected by the performance and behavior of key individuals and institutions. The loss of two or three key people can mean the end of crucial programs. At the same time, small institutions can increase the potential impact of key decisions and the influence of research leaders.

Small countries are also more susceptible to environmental hazards and disasters: on a small island a cyclone may devastate a valuable export crop, wiping out a year's foreign exchange earnings in a single night. More important still, minor miscalculations in policy and strategy can have major repercussions on the performance and survival of key institutions. This increases the importance of management decision-making, and of the information that supports it. Flexibility, and a rapid response to internal or external change, become all-important for survival.

Dependence upon external resources and agencies is another source of vulnerability. Foreign aid helps small countries compensate for their small income base, but it also makes them highly vulnerable to changes in the policies of donor agencies. It is often difficult for national policy to direct and manage donor inputs in accordance with national objectives, despite the good intentions of both parties to work together. On their side, donors care less about withdrawing their assistance to small countries, since the political fall-out from withdrawal is likely to be less. In some agencies there has also been a somewhat complacent assumption in recent years, that small countries do not need a national research capacity. Instead, they should "beg, borrow, or steal" technology from their larger neighbors.

This is a dangerous oversimplification and a chronic underassessment of the true requirements of adaptive research. A national research capacity is necessary in small countries if they are to select the useful technologies that can be adapted to national conditions. Crucial decisions on agricultural development and natural resource policy will require inputs from local scientists with a national perspective. National research institutions in small countries are both essential and feasible. The size constraint can be overcome, but it requires new approaches to policy and the organization of research to mobilize institutional resources in more flexible and diverse ways. Some options and strategies to accomplish this are presented in this book.

CAN SMALL COUNTRIES HAVE VIABLE NATIONAL AGRICULTURAL RESEARCH?

The distinguishing characteristic of a small country is that the scale of its research system will remain small, even under the best of circumstances. The demand for research, however, remains large, covering the same broad range of problems that confront larger countries. Conventional wisdom dictates that a small country should narrow its scope to cover only those few areas that it can handle well. When faced with the realities of policy, development, and the environment, however, the proverbial common wisdom offers little practical guidance as to what research leaders and policymakers should do. They cannot build bigger research institutions to do more, nor can they afford to do without the results of research or without information and advice on the best way to develop the agricultural resource base.

Similarly, small countries should not be left out of global agendas on agricultural science and the environment. The implications of sustainability, developments in biotechnology, and agricultural diversification for national priorities are, if anything, even more pressing and challenging for small countries. All countries, rich or poor, large or small, need an institutionalized capacity to generate or develop, interpret, and adapt scientific knowledge and new technologies to meet the objectives of national development. Building an effective and sustainable agricultural research capacity requires that research leaders and policymakers consider several strategic questions (see Table 1.1).

In small countries, the answers to these questions are more difficult and somewhat different than they are for larger countries. Because there are stringent limits on the human and financial resources that can be invested in research, small countries may not

Table 1.1. Strategic Questions for Research Policy and Organization

Questions	Small-Country Conditions
How much can a country invest in research?	• Funding as % of AgGDP in small countries is already higher on average than in larger countries; further increases are unlikely • External sources of funding are not growing • Trained human resources in science and administration are scarce and difficult to retain
How can research capacity be organized and institutionalized?	• Few national research institutions are big enough to cover the breadth and scope of the problems • Many dispersed activities in projects, NGOs, and producer associations • Difficulty in capturing and applying relevant information and technology from outside and difficulty storing and using information about resources and innovative technologies being tested in-country
What are the key functions to be performed?	• Experimentation • Managing information • Coordination • Policy advice • Regulation • Linkages
How can a realistic scope be set and sustained?	• Commodity domains • Natural resource problems • Socioeconomics, post-harvest, and marketing themes • Diverse institutions doing research within these domains
How can the most be made of technology, information, and resources from outside?	• Regional partnerships • Networking • International research centers • Donor projects • International agencies

be able to build the large research institutions with the full-fledged adaptive research programs and studies that are traditionally defined as the functions of NARS.[5] Instead, small countries need to devote greater efforts to studying their production systems and natural resource base in order to match technologies to their circumstances. They need more emphasis on information as a tool in policy setting. Finally, they will need greater emphasis on links to external sources of technology and information and on the coordination of research within the country by a wide range of institutional actors, some of whom stand outside the core research system funded by government.

FAILURE OF EXISTING STRATEGIES

The last two decades witnessed a considerable growth in research capacity in developing countries, including small countries. Major donor projects to build the institutional capacity of NARS were implemented in many small countries. While donor funding often plugged useful gaps, it too was subject to dramatic fluctuations. Evidence from our case studies confirmed time and again that small countries can ill afford to base the size of their research system on the "boom-and-bust" cycles of donor funding.

Global studies show that the level of resources in small countries, as measured by research-intensity ratios of funding per researcher or number of researchers per hectare, compares favorably with other developing countries (Pardey, Roseboom, and Anderson 1991). However, the small scale of the research systems and agricultural sectors tends to distort the picture. The final result in many cases is a small number of researchers concentrated in a single institution with a very broad mandate, trying to conduct commodity improvement programs with resources that are inadequate or ill suited to the task.

In several cases the sustainability of NARS in small developing countries has been called into question (Eicher 1988). "Concentrate resources and narrow the focus" are two common pieces of advice that have been widely offered. However, these solutions may be unrealistic and too simplistic. Perhaps the answer is not for small countries to try to do less than the NARS in larger countries, but to do things differently.

To succeed, a small-country research system must innovate, not always by producing new technologies, but in the approaches it uses to seek, organize, and apply these technologies. The NARS in small countries need to do things differently, and innovatively, so that the appropriate technology and information reaches farmers and policymakers. A strategy for research in small countries needs to reflect the following conditions that are specific to small countries:

- Some of the organizational options will be different.
- A different set of functions may be emphasized.
- Setting an appropriate scope of research will involve a division of labor based on the comparative advantages of different institutions.

- Small research institutions may require a different type of program organization and scientific skills.
- Research priority-setting will incorporate a broader set of factors.
- National research planning will increasingly be done within a regional context.

To address the difficult task of research planning in a small developing country, we have developed what we call a *portfolio approach*. In this approach, the entire scope of research, together with the technological and policy environments, is examined to determine the comparative advantages of different research institutions in addressing the different functions of the research system and in conducting research in a given domain. Policymakers and research leaders can use this approach to allocate responsibilities and resources to institutions across the various domains of research.

This approach and its elements are the core around which this book is structured. Further chapters look in more detail at *scale, scope, linkages,* and *information flows,* drawing on examples that demonstrate how this approach may be applied.

THE NATIONAL RESEARCH PORTFOLIO: A NEW STRATEGY

The research portfolio approach is designed to accommodate three features of research management in small countries: first, their need to rely on a diverse set of institutions to provide needed technologies and information; second, the difficulties that they encounter in narrowing the scope of their research, given the increasing demand for new technologies and information; and third, their need for a research capacity that is able to handle the broad set of functions required of a small system.

The approach presented here is adapted for use in small national research systems seeking to maximize national food and income security. Other versions of the approach have been proposed as a tool for improving the efficiency of large industrial research and development organizations (Remoortere and Boer 1992). Applying the portfolio approach entails assessing varying levels of risk and institutional performance. It also presents opportunities for partnerships, joint ventures, and mergers.

The portfolio approach enables research planners to identify the comparative advantage of different research organizations for different functions. It also allows priorities to be set on the basis of a broader set of factors. In the past, the key factors have been the importance of the problem, the probability of success, and the expected rate of return to research investments. The approach we propose here also incorporates the availability of external information and technology. The management of information, in particular, is seen as a central function of the national research system in small countries (see Box 1.2).

> **Box 1.2. What is the Research Portfolio Approach?**
>
> The research portfolio approach is a strategic planning tool that a NARS can use to make the most of institutional diversity under severe scale limitations. A portfolio approach is useful for identifying institutional comparative advantages and ways of managing research spillovers, as well as for deriving economies of scope among programs, institutions, and countries. It is used to
>
> - vary levels of research investment
> - select NARS functions that can be performed well and in a cost-effective way
> - identify the most appropriate institution within or outside the NARS to perform a function
> - develop a unified research policy that coordinates and establishes shared objectives among research activities and institutions
>
> *Source:* Eyzaguirre (1991).

TWO KEY CONCEPTS

We use two concepts to help us understand the size constraint in the organization and management of national research systems in small countries. The first is the concept of *scale*. This is defined as the institutionalized research capacity of a national system. It is a combination of a system's human and financial resources, its knowledge base, and its infrastructure. The second concept is that of *scope*. This refers to the research agenda of a system, including the set of research topics and objectives to which the national institutions are committed. *Scope* has two dimensions: the range of research programs, meaning the commodities and topics covered, and the level of research, meaning whether it is strategic, applied, or adaptive. Matching the scope of research with the scale of available resources is a key management challenge in a small country.

FOUR STEPS IN THE APPROACH

In Chapter 3, we present a strategy that allows research managers and policymakers to transform broad research scope into a manageable research portfolio. It entails several steps or tasks (Figure 1.2) and requires research domains and research functions to be clearly identified. The first two steps are discussed further in Chapter 2; Chapter 3 focuses on the third and fourth steps. The steps are listed below.

1. Unify policy

First, members of the various components of the research system are brought together to build a consensus on national policy goals for agricultural research. Building a national research system that increases the scale of the research effort depends on coordinating existing research institutions and units rather than building bigger organizations. This must be done at the policy level.

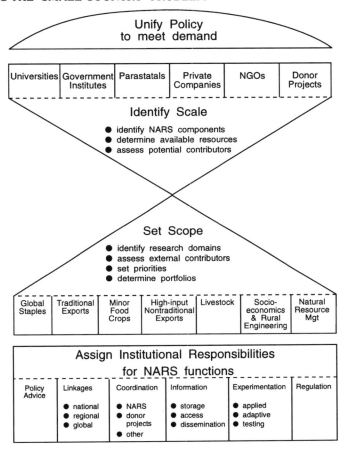

Figure 1.2. Elements of the research portfolio approach

2. Identify scale

Second, much analysis of research looks at only part of the national research effort, proposing models that seek to build the NARS on a single stem (public, private, or parastatal). Our portfolio approach, however, builds upon institutional diversity where *all* potential contributors to agricultural research are identified and mobilized.

3. Set scope

Third, when core NARS institutions can adjust their scope to establish complementarity with other actors, resources can be freed to allow work in areas that these other agencies would not cover. Allocating responsibility for export-oriented commodity research to parastatals or private producer groups allows public-sector research to focus on issues of food security and natural resources.

4. Assign institutional responsibilities

Ensuring that all the research that is needed is also implemented requires an agreed-upon division of labor between institutes and organizations. Government research organizations still have a central role to play in coordinating and setting priorities so that research activities, even those outside their direct management, complement one another and work to the long-term benefit of the country. Examining the entire scope of research allows the comparative research advantages of different institutions to be identified, and responsibilities and resources to be allocated.

One of the things that is not different about small countries is the broad demand for research. Research managers face many competing demands for their services and for scarce resources. The problem is to manage the broad scope that results from this demand. One way is to break the potential scope of research into discrete domains, each defined according to criteria based on the intensity of technology generation, the intensity of technology and information flows, and the type and focus of research topics. Based on our information on the scope of research in small countries, we were able to group topics and themes into seven broad research domains for analytical purposes: global staples, traditional exports, minor and traditional food crops, high-value nontraditional exports, livestock, socioeconomics and rural engineering, and natural resource management.

Segmenting the national scope of research into domains enables policymakers to consider the institutions and resources involved as a type of national investment portfolio in agricultural research and development. Any portfolio will have varying levels of risk and institutional performance and will present opportunities for partnerships, joint ventures, and mergers. The framework we propose is aimed at research leaders and policymakers in the public sector who are involved in national agricultural research planning. Our goal is to provide a framework whereby the national research portfolio can be managed to ensure long-term food and income security in a developing country.

To conduct applied or adaptive research covering all the domains of the research portfolio described above would be beyond the capacity of most developing countries, not to mention a small one. For this reason, the portfolio approach emphasizes a broader set of NARS functions. Traditionally, the function of a NARS has been understood largely as *experimentation*. Where natural resource management is concerned, the diagnostic function is crucial. The portfolio approach recognizes the following set of NARS functions in addition to experimentation and diagnosis: *policy advice, managing linkages, coordination, managing information,* and *regulation.* Applying the portfolio approach allows functions to be shared among institutions, as well as across domains.

The following chapters discuss how the institutions within national research systems can allocate and share responsibilities and functions to meet the broad demands of their research scope. The solution is not organizational growth, but innovative policies and management mechanisms to make the most of institutional diversity, to live with functional complexity, and to strive for flexibility. The findings of our studies show these to be the key elements in making agricultural research systems in small countries feasible. The findings from small countries are also relevant to other developing countries facing increasing demand for research outputs with dwindling resources.

Notes

1. Twenty-five of the 47 countries that the United Nations has classified in the least-developed group are small countries by our definition (UNCTAD Secretariat 1992).
2. The disadvantages and problems of small island states have been well reviewed by Dolman (1985). They include
    a. few economies of scale;
    b. high degree of vulnerability to natural disasters and external shocks;
    c. narrow resource base;
    d. distant markets;
    e. poor market access for both exports and capital;
    f. limited independence in policy-making;
    g. limited skilled manpower.
3. A full discussion of the criteria used to identify small countries in the ISNAR study is provided by Eyzaguirre (1992).
4. The population pressure on agricultural land, measured in terms of agricultural population per km^2 of agricultural land, ranges from 343 persons/km^2 in Bhutan to 222 persons/km^2 in Burundi to 145 persons/km^2 in El Salvador. Rural poverty exacerbates resource degradation because rural people naturally place their own short-term survival over the long-term conservation of the environment.
5. ISNAR and the Consultative Group on International Agricultural Research (CGIAR) include crops, livestock, aquaculture, and the management of natural resources (including forestry, agroforestry, and fisheries) as part of "agriculture". The national research institutions with specific mandates to address problems in these sectors make up the NARS.

2 Working with Small-Scale and Diverse Institutions

INTRODUCTION

In Chapter 1, we referred to the diseconomies of institutional development in small countries. These are countries that can build neither more nor bigger organizations to confront the full range of development problems that they face. We can argue that there are not enough scientists to meet the demands of agriculture, forestry, and fisheries, but this constraint is not unique to agriculture, nor is it even necessarily the most severe problem the country faces. Trained technical professionals are in short supply in health, education, and transport as well. The choices for policymakers are not easy: each scientist trained in agriculture means one less doctor or teacher. Not only is there an absolute shortage in the quantity of training available, but its quality too may be severely limited. Very few of the small countries in Africa or the South Pacific, for instance, can train their professionals to the level needed to carry out the complex tasks required for conducting advanced agricultural research.

This chapter discusses how to build viable research systems in situations where research organizations work with few researchers and fragile, stagnant, or declining funding bases. Our studies suggest that an effective agricultural research system in a small country cannot be based on a single institution, public or private. Research leaders must focus on making diversity work for them—through attention to policy and coordination.

We begin by highlighting the first two steps in the portfolio approach: identifying the scale of research available to a country and formulating a coherent research policy that identifies the comparative advantage of the institutions that contribute to research (Figure 2.1).

Table 2.1 shows data on the number and type of institutions involved in agricultural research in 46 small countries.[1] Table 2.2 presents information on the scale of resources devoted to different types of institutions in 24 of these countries, for which both personnel and financial data were available.[2]

The tables reveal considerable institutional diversity. In addition, there is some disparity between the distribution of researchers and the allocation of financial resources. Publicly administered government research institutions are by far the most numerous type of institution, and they account for over 75% of all researchers.

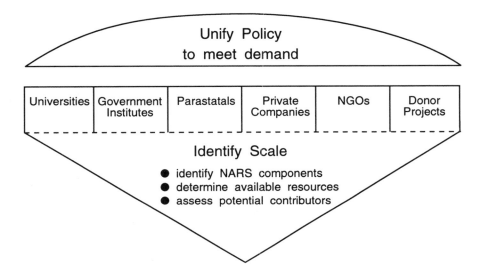

Figure 2.1. Policy and scale components in the portfolio approach

Table 2.1. Number and Types of Research Institutions in 46 Small Developing Countries

Type of Institution	Core Institutions	Supporting Institutions	Total
Agricultural Research Councils	10		10
Research Foundations	3	1	4
French Tropical Research Institutes		7	7
Ministries	5	2	7
Government Agricultural			
Research Institutions	76	32	108
Government Research Institutes	5	8	13
Agribusiness			
Multinational	4		4
National	2	12	14
Parastatal Organizations	14	16	30
Nongovernment Organizations		1	1
Regional Organizations		2	2
Regional Research Organizations	1	3	4
Universities			
Regional		3	3
National	2	17	19
Total	*122*	*104*	*226*

Note: For an explanation of the types of research institutions, refer to Annex 1.

Table 2.2. Institutions and Their Resources in 24 Small Countries

Type of Institution	Number of Institutions[1]	Number of Researchers[2]	Total Annual Expenditure (US$)[3]	Expenditure/ Researcher (US$)
Multinational Agribusiness	2	9	2,800,000	315,000
Government Agricultural Research Institutions	39	894	50,544,616	88,352
Parastatal Organizations	12	161	10,411,481	64,191
Nonagricultural Government Research Institutes	6	58	2,270,596	58,183
Research Foundations	1	29	817,910	28,204
National Universities with Faculties of Agriculture	3	108	2,231,580	20,663
Total	63	1,259	69,076,183	574,593

Note: Data are from years 1988 to 1992.

1. This is from a sample of 63 research organizations in Barbados, Benin, Botswana, Burundi, Cape Verde, Chad, Fiji, Gambia, Guyana, Honduras, Jamaica, Laos, Lesotho, Liberia, Mauritania, Mauritius, Papua New Guinea, Paraguay, Rwanda, Sierra Leone, St. Vincent, Swaziland, and Togo. Selection was based on availability of data on both expenditures and personnel.

2. Distribution of data by year: 1988 = 11; 1989 = 11; 1990 = 24; 1991 = 15; 1992 = 2 (see Annex 3).

3. Expenditures are in local currency (LCU) converted to constant 1987 US dollars at the average market exchange rate.

COMPONENTS OF THE NATIONAL RESEARCH SYSTEM

The core[3] national research capacity includes government research organizations in agriculture and natural resources (with varying degrees of administrative autonomy), parastatals and single-commodity boards or institutes, research foundations, research units in private companies, and some universities (see Annex 2).

Besides these mandated institutions, national systems can make use of the ad hoc or noninstitutionalized research activities that often take place in donor development projects, nongovernmental organizations (NGOs), producers' associations, and universities. While they may not be part of a long-term national plan or program of agricultural research, these activities are important sources of knowledge and experience in areas that government research organizations cannot cover.

GOVERNMENT ORGANIZATIONS

The most common type of government research institution is found within ministries of agriculture, natural resources, or primary industries. Research is done by departments, divisions, directorates, or centers within ministries, and the research units are often part of larger administrative units dealing with crop and livestock development. They range in size from one or two researchers (in the Seychelles) to around 60 (in Benin), and their budgets rarely exceed US$ 1 million (constant 1987 US$).

Several countries have more than one such institution. Togo, with its four public-sector institutions all working in agricultural research, is a good example. In recent years, many countries have attempted to get around the fragmentation of their research effort by consolidating their public research units into a single organization.

This brings us to the second type of government research institution, the autonomous or semiautonomous public-sector research institute, which often has its origins in the consolidation of previously separate research units. Typical examples are the National Agricultural Research Institute (NARI) of Guyana, the Institut des Sciences Agronomiques de Burundi (ISABU), or the Instituto de Investigación Agropecuaria de Panamá (IDIAP). These institutes are generally larger than research units within ministries, averaging slightly more than 40 researchers per institute.[4] A feature of these institutes is their broad scope, which typically stretches their resources, sometimes beyond the point at which critical mass can be maintained. One positive feature is that concentrating resources and research activities in a single institute raises the visibility of research and increases its potential to influence policy decisions.

These two kinds of government research institutions are clearly the dominant research components in small developing countries. A distinctive feature of both institutional types is their multifunctionality. Research is usually only one of their responsibilities and staff often deal with seed production, extension, and farmer training, and provide analytical services as well. This means that clearly differentiated research activities are difficult to discern and manage. Nevertheless, given the greater relative importance of the coordination and policy advisory functions of research in small countries, these institutions are likely to remain at the core of small-country national research systems. In many such countries, however, a period of further organizational change is needed to convert them into institutes that can conduct research with a systems perspective and with sufficient stature to advise governments on technological and policy options.

PARASTATAL ORGANIZATIONS

In many countries, large and small, research on traditional export commodities is organized in separate parastatal organizations or commodity boards or institutes. A parastatal is any organization mandated by government to operate as a private agency in a scope or sphere that falls within the public domain. Public functions such as levying and spending taxes and duties on agricultural exports are often transferred to the parastatal by government. Parastatals are managed as private entities with statutory authority to apply a tax to commodity exports and to use it for agricultural R&D.

Typically, government retains a strong interest in the activities of the parastatal and is represented on the governing board. Compared with government organizations, parastatals tend to be better funded (especially with regard to operational funds) and have more efficient management, better links with producers, and a narrower focus on key cash crops.

Papua New Guinea is an interesting case where a major investment in parastatals was made during the 1980s (see Box 2.1). In most countries, parastatals date back to independence when many commercial plantation complexes were nationalized. Examples from the sugar industry include the Fiji Sugar Corporation, Caroni Ltd (Trinidad and Tobago), and the Guyana Sugar Corporation, which all grew out of foreign companies nationalized by government. Each of these maintains research departments that are important components of their respective national agricultural research systems.

Box 2.1. Parastatals in Papua New Guinea

In Papua New Guinea, four parastatals were created in the mid-1980s to conduct research on the country's traditional export crops: coconut, coffee, cocoa, oil palm, and sugarcane (Sitapai 1992). They are jointly owned by the government and the commodity boards and are financed by a tax on exports together with direct grants from the Ministry of Agriculture and Livestock.

Their share in the national research system is substantial. In 1992, 50% of the country's 62 agricultural researchers were employed by the parastatals, and their research expenditures were 51% of the national research budget. This compares to the Ministry of Agriculture's 37% share of the national research budget.

Parastatals can diversify their research interests and thereby help diversify the national economy: research at Caroni is now concentrated as much on nonsugar commodities as on sugarcane, supporting the national drive to develop nontraditional exports.

Research units in commodity-based parastatals have advantages over their counterparts in the government sector. Key among these is the ability to raise and manage their own funds and staff, independent of the bureaucratic norms of the civil service. They have traditionally fared better in terms of stable funding, mainly due to their regular income from a tax on commodity exports. However, commodity parastatals are not as well funded as they were in the mid-1980s: the average level of funding per researcher is no longer higher than it is in government research organizations. Some of the apparent decline may be explained by the fact that parastatal boards usually have lower expenditures for administration and donor projects, both of which tend to inflate total expenditures in government institutions. But most of it is probably due to the declining terms of trade for the traditional export crops of developing countries, which reduces the amount of money raised by taxing exports.

This brings us to the major limitation of commodity parastatals in many countries, namely the small size of the commodity subsectors for which they are responsible. The income from a tax on a single export commodity is rarely sufficient to fund a fully fledged research program. Policymakers reviewing the possibility of establishing a parastatal need to consider the size and stability (yield and price) of the commodity concerned to determine whether privately financed research is feasible.

Our study revealed several mechanisms by which parastatals fund research without assuming responsibility for executing it. Because of declining revenues, research managers in many parastatals are increasingly pursuing these options to ensure that their research needs are met. This trend is particularly marked in the Caribbean, where traditional agricultural exports have been in decline for many years. The parastatals are calling on governments to assume more responsibility for funding research, allowing the parastatals to focus on the nonresearch aspects of their mandates. In some cases, parastatals are contracting their research out to other organizations instead of creating their own research capacity. The Citrus Growers Association of Jamaica contracts short-term research to consultants and to the Sugar Industry Research Institute (SIRI), while analytical work and longer-term research is contracted to the country's Ministry of Agriculture. Also in Jamaica, the Coffee Industry Board relies heavily on technology developed in other coffee producing countries and subcontracts much of its research to the Ministry of Agriculture, CARDI (a regional research organization), and to the University of the West Indies (Reid 1993).

This approach does not always work, despite the obvious advantages of pooling resources. The contracted agency may have difficulty retaining the necessary staff or managing the research programs required by producers. Or it may not have as thorough an understanding of the problems faced by farmers. While short-term consultants may have proven experience, they seldom have sufficient time to devote to solutions, nor can they delve too deeply into the underlying causes of production constraints. The experience of the Swaziland Cotton Board is salutary (see Box 2.2).

RESEARCH FOUNDATIONS

Research foundations are private, nonprofit organizations with mandates to work in areas of the public good. They operate in the private sector under independent management but are oriented towards national development goals. They are intended to stimulate private-sector research and development in areas of strategic national interest. In both Jamaica and Honduras they support research on high-value nontraditional exports. In El Salvador, the Fundación Salvadoreña para el Desarrollo (FUSADES) was set up to promote research and development to meet the needs of resource-poor farmers.

Foundations typically receive public support in the form of tax exemptions, grants, and donor funding but are mandated to seek private funding and can enter into contracts with other organizations, public or private, for services. They may fund, execute, or coordinate research. Their policies are established by independent boards normally consisting of senior public officials and representatives from the private sector. Foundations are thus mechanisms for promoting joint public/private ventures to meet national development objectives. The underlying premise is that despite the desirability of private-sector involvement, in most developing countries this will not materialize without some support from the public purse (see Falconi 1993).

**Box 2.2. Contracting Parastatal Research
to the Public Sector in Swaziland**

The Swaziland Cotton Board (SCB) funds some recurrent expenditures at the research
stations of the Ministry of Agriculture's Agricultural Research Division (ARD). The Swazi
government matches the board's funds by providing personnel, office space, and infrastructure. There are presently positions in ARD for a cotton entomologist and a cotton breeder.
Funds for these positions are raised from a levy on cotton producers. Priorities for research
are set annually, at a meeting of the Cotton Research Committee; a second meeting at the
end of the season assesses past results and progress. For much of its technology and
information, the SCB maintains strong links to research in the Republic of South Africa.
These arrangements appear favorable to the cotton industry, which provides funds to
government and in theory receives the benefits of research it could not otherwise afford.
However, research on cotton breeding over the past 10 years has been plagued with
problems resulting from the small scale of the country's research institutions.

During the 1980s, the quality of the cotton crop in Swaziland declined, prices fell, and
research was needed to identify and maintain high-quality seed and fiber, mainly through
breeding. In 1983, a researcher was sent abroad for MSc training. On his return, however,
he was appointed director of all public research in the country, which left him with little
time for detailed research on cotton. In 1985, therefore, the SCB contracted a South African
cotton company for the part-time loan of their breeder. Between 1985 and 1988, the
borrowed breeder was assisted by a trainee breeder, who was later sent abroad for MSc
training. In 1989 the trainee, now qualified, left government service and the SCB was back
to relying on a previous research supplier, the Tobacco and Cotton Research Institute in
South Africa. The board then obtained the services of a short-term expatriate breeder from
the United Kingdom, but no counterpart was made available by the ministry. In 1990, the
board appointed its own trainee cotton breeder and the Ministry of Agriculture agreed to
recruit another cotton breeder. Both needed training abroad. By the end of 1991, the
expatriate breeder was nearing the end of his contract, there was still no local replacement,
and the board was continuing to plan for two local cotton breeders.

This case illustrates some of the problems encountered by a small country that seeks to
build a critical mass for research on a commodity, even when research funds and long-term
commitment from the industry are available.

Source: SCB Annual Reports 1983–1991, and discussions with SCB staff.

The growing popularity of research foundations as a way of securing greater private
participation in national research and development may prove short-lived. Some
studies have argued that, though nominally private, research foundations are in fact as
dependent on donor financing as some public-sector research organizations (Sarles
1990). Most of the foundations in the countries studied by ISNAR had been established through donor funding. In our sample, only Fundación Hondureña de Investigación Agrícola (FHIA) received significant private-sector financing (see Box 2.3).

In general, foundations have arisen in those countries with an important agricultural export sector that is seeking to diversify. In Jamaica, the commodity boards that
had been the major source of private-sector support for research were no longer in a

---

Box 2.3. Stimulating Private-Sector Research in Honduras: The Role of FHIA

The Fundación Hondureña de Investigación Agrícola (FHIA) was created in 1984 to promote agricultural exports, increase food production for internal consumption, promote agricultural diversification, and decentralize and privatize aspects of agricultural research. The United States Agency for International Development (USAID) provided an initial grant of US$ 20 million to establish the foundation. It also assisted in setting it up. United Brands, a transnational corporation, provided the major private contribution by donating its research complex at La Lima to the new institution. Seven years after initiating activities as an innovative joint "project" of government and USAID, FHIA is well on its way to becoming fully privatized. However, it still receives government assistance and conforms to public-sector policy and strategy objectives.

FHIA's 1992 budget was approximately US$ 3 million, of which 60% was from USAID, 10% from laboratory and diagnostic services, and the remainder from various projects and grants. The 51 scientists conduct research and transfer activities on selected export and food crops, including bananas, plantains, cacao, black pepper, mangoes, and palm hearts. While this research is mostly applied and adaptive, FHIA is among the world's leaders in research on bananas and plantains.

FHIA's effectiveness in applying private-sector approaches to public research is based on its ability to attract and retain well-qualified staff, its avoidance of any undue political influence on its operations, the provision of appropriate logistical support to technical staff, and limiting bureaucracy.

Source: Fernandez (1992).

---

position to maintain their independent research facilities as prices for the country's traditional export commodities fell. Under these circumstances, a foundation, with its mix of public and private funding, appeared to be a more flexible mechanism for providing research support. It could also stimulate the sharing and consolidation of research facilities for several commodities supported in the past by independent boards. In new research areas with a relatively high risk of failure, such as initial research on high-value nontraditional export crops, foundations are also a useful mechanism for persuading private producers to become involved. This is a key role for the new Jamaica Agricultural Development Foundation (JAFD), a research advisory and funding mechanism that will allocate approximately US$ 1 million each year to research projects in the Jamaican Ministry of Agriculture, the private sector, the University of the West Indies, and overseas universities (Wilson 1992).

At present, foundations are found mainly in Latin America and the Caribbean. Whether this model will be exported to other parts of the world depends on the willingness of donors to invest public funds in private organizations and the degree to which this "seed money" attracts private-sector involvement. As we have seen, foundations have been criticized for purporting to be private initiatives when they are really largely dependent on donor funding, but their record in tapping new funding sources does not yet appear very successful. Foundations are often regarded as just one more type of research organization drinking from a shrinking pool of public funds.

The foundation may yet provide that elusive common ground on which public and private interests can invest jointly in research of national importance, with a profitable return for both investors and consumers, but its future is on the line.

PRIVATE COMPANIES

A number of multinational agribusinesses or transnational corporations have established research institutions in several small developing countries. Fruit exporting multinationals, such as United Brands and Standard Brands, have research units in Central America. Companies such as Dole, Del Monte, Birds Eye, and Libby are active in research and technology transfer from Latin America to Africa. The countries in our sample in which these firms are active are primarily in Central America. Swaziland and some of the South Pacific nations also have important research units attached to private firms. Their research is focused on fresh and processed fruits and vegetables for export. Other industrial firms such as Unilever are active in research and development of tropical oil crops, principally oil palms and coconuts, in areas ranging from West Africa to the South Pacific. The rubber industry multinational, Firestone, had established a major research and development center in Liberia, but this is now in sad disarray owing to the civil war in that country.

In most small countries, institutionalized research carried out by agribusiness companies takes the form of small research units of two to five scientists working mainly on production constraints, addressing problems in plant protection as they arise, and testing improved varieties. These teams are well funded on the basis of expenditures per researcher. In Honduras, the two largest research units, belonging to Chiquita Brands and Standard Fruit, together employ 15 researchers (around 9% of total scientists in the Honduran national system), but they consume 35% of national expenditures on agricultural research.

Many multinational agribusinesses no longer manage research units themselves but support or contribute to agricultural research by contracting research services from government institutions. Others, such as Cadbury, the chocolate manufacturer, provide direct grants, funding cocoa research and conservation of genetic resources in the Caribbean and West Africa. In Sierra Leone and Mauritius, the British–American Tobacco Company is active in supporting tobacco research and development.

These firms have so far focused most of their research on the traditional export commodities that have provided most of their profits. But they have increasingly been broadening their scope to include high-value nontraditional exports. In Central America, for example, the "banana companies" are major actors in the shift towards newer crops such as grapefruit and melons (Rojas 1993).

A few private-sector agricultural input companies, among them Pioneer, Northrup King, and the chemical multinational Ciba-Geigy, provide financial and technical support to public-sector research institutions working on global staples such as maize and sorghum. Botswana has been able to attract investments from such companies for maize research and seed production. However, since internal markets are small, the opportunities for attracting this kind of investment are not great in small countries. Only when a small country is seen as an entry point for diffusing technology and information to a larger region does it attract investment on a larger scale.

Participation by multinationals in agricultural research and development is a thorny issue for some small developing countries. The large banana multinationals in Central America often used to disregard national policies and interests, while foreign palm oil and coconut plantations in the South Pacific were not renowned for their concern for local welfare and national development. The sheer size and economic power of some multinationals can dwarf the economies of the small countries in which they operate. It is understandable, therefore, that small countries approach relationships with multinationals with some trepidation. Nonetheless, some multinationals have established fine research complexes to support commodities of strategic importance to a country's development. We can point to the research complex for bananas established by United Fruit in Honduras, Firestone's Rubber Research Institute in Liberia, and Unilever's coconut research facilities in the Solomon Islands as examples of investment by private multinationals in applied national research.

Developments in areas such as biotechnology (which means that some types of strategic research need no longer be done locally), the growing competition among producers, and the speed with which technology transfer and information exchange take place have changed the way research is done by multinationals, making it more difficult for them to appropriate the results of their research. Multinationals have therefore been divesting their research interests to national nonprofit management structures. Thus FHIA inherited the United Fruit research complex at La Lima (see Box 2.4), while the Liberian Rubber Research Institute took over Firestone's research facilities. Private firms will continue to conduct *maintenance research,* but this can be done by very small teams of researchers: often no more than two attached to the production companies.

The experience with research by multinationals in small developing countries offers a clear message for policymakers and national research leaders. Where such research is established, the public and parastatal sectors can build activities that complement these efforts. Through their links with private companies, these sectors can extend the results of private-sector research into other domains and provide policy guidance so that private-sector research contributes to national development goals.

Favorable public policies and initial public investment are important conditions for securing private participation in research (Falconi 1993). Conversely, private investments and partnerships are a good way for public research to gain additional resources. Where private companies have organized their own research capacity, this allows public institutions to adjust their activities to tackle crucial areas that the private sector is less likely to cover.

Box 2.4. Banana Research by Multinationals in Honduras

In the past, United Fruit bore most of the costs of developing technologies for the banana industry, on the assumption that it could recover the investment through its sizable share of the world market (estimated at 60% to 70% during the 1950s and 1960s). This share has fallen in the past 20 years, with the result that some of the incentives for the company to invest in research have been lost. The relative ease with which new production technologies in bananas can be appropriated by competitors is also a critical factor here.

The type of long-term investment made by the company when it developed the world-class facilities for applied research in Honduras is no longer seen as justified. There is still a need for applied and, indeed, strategic research, but this will increasingly be conducted in the public domain, by national institutes such as FHIA and international centers such as the International Institute of Tropical Agriculture (IITA). There will be benefits to Honduras from the industry's continuing research on production and postharvest problems, but the industry is unlikely ever again to invest in any long-term research capacity that can be used outside the immediate needs of the banana export sector.

Source: Contreras (1992).

UNIVERSITIES

Many small countries cannot afford universities with faculties of agriculture that contribute to research. Where such universities do exist, however, the national research system tends to be stronger. A university can provide a link to the global science community that is useful for maintaining the credibility and prestige of the research profession (see Box 2.5). Given the potential for isolation and the need for many researchers to spend time developing downstream technologies, the access to a university helps to overcome the feeling of scientific isolation. Universities also are a focal point for scientific exchanges and visits, providing an additional avenue for acquiring external knowledge. The presence of a university has a positive effect on the scientific climate in general, especially on access to information. Libraries within small research institutes are notoriously expensive and difficult to maintain; a university can provide a large central library to service many smaller research units and institutions.

Several of the ISNAR small-country case studies highlighted the problem of keeping scientists in the system, and universities can provide a larger pool of nationals to replace the staff who leave the research system. The ability to draw on university faculty members to complement scientific staff and the steady supply of new graduates can help research institutes mitigate the effects of staff attrition and budget cuts that are such a common feature of small national systems. This has been the case in Sierra Leone (Dahniya 1993).

There are two domains in which universities may have a comparative advantage: natural resource management and socioeconomic and policy studies. This is because research requiring a longer time frame, and research that is information-based as

---

Box 2.5. Linkages with Universities

In some countries, research institutes are based within the university and draw upon university faculty members, who divide their time between research and teaching. Examples of university-based institutes with agricultural and natural resource mandates are the Centro Nacional de Investigación Forestal Aplicada (CENIFA) in Honduras, the National Institute of Development Research and Documentation (NIR) in Botswana, and the Institute of Marine Biology and Oceanography (IMBO) in Sierra Leone.

Regional universities with schools or faculties of agriculture have made notable contributions to research. The Faculty of Agriculture of the University of the West Indies has been a center of excellence in tropical agriculture for many years. The University of the South Pacific's School of Agriculture is a major contributor to research in the small pacific island nations.

In other countries, universities, such as Njala University College in Sierra Leone and the University of Technology in Papua New Guinea, were specifically established to serve agriculture and the natural resource sector. The Pan-American School of Agriculture in Honduras is a notable example of a private agricultural university that has made important contributions to agricultural research in the region.

In Sierra Leone, Botswana, Lesotho, and Namibia, the crop and livestock research institutions share a campus with the agricultural college, increasing the opportunities for interaction and shared facilities. In Mauritius, the sugar research institute, the ministry of agriculture, and the Food and Agricultural Research Council are all within a five-minute walk of the university, giving them a variety of options for sharing resources and coordinating their activities.

---

opposed to technology-based, is often best done by universities, and the capacity to conduct this type of research is often greater in universities than in other institutions. For example, the School of Agriculture at the University of Mauritius has conducted a 15-year study of leucaena and its impact on soil fertility and biomass.

While research in small countries will be called on in the future to move closer to users and clients and to perform other development functions besides experimentation, the performance of the NARS depends upon the staff and the institutions maintaining a solid culture of science for development. The presence of a university is a positive factor for sustaining this science culture in a NARS. We do not intend this to be an argument for a university in every country, which would only exacerbate the scale constraint in many small countries. What we are saying is that where universities exist, the environment and the potential resources that research organizations can tap are much richer.

NONGOVERNMENTAL ORGANIZATIONS

The nongovernmental organizations at work in the small countries studied by ISNAR varied considerably in their mandate, size, and type of operation (service-oriented or policy-oriented). Many of them had activities that directly affected the agricultural sector, yet agriculture was not their major concern. Some, however, were focused on agriculture, often on issues of natural resource management and conservation. Examples in this category are the Caribbean Natural Resources Institute in Saint Lucia and Thusano Lefasheng in Botswana. Botswana has a large NGO sector whose leading actor, the Rural Industries Innovation Center, has played a major role in the domain of rural engineering and food technology.

Issues. The Overseas Development Institute (ODI) has completed a global study of the role of NGOs in agricultural development (Farrington et al. 1993). Their cases from Africa, Asia, and Latin America demonstrate the roles of NGOs as contributors, executors, and clients of agricultural research. Where the institutional scale of formal research and development remains small, as in the countries we have selected, the role of NGOs assumes greater relative importance. A major advantage of NGOs is their ability to mobilize external funding from private voluntary contributions as well as public sources. They also have close links with the rural poor, have a strong concern for their welfare and interests, and are good at involving local communities in self-help and development efforts. These characteristics make them important actors in research that is local and community based.

However, access to the research services and outputs of government research organizations is a condition for the success of NGOs. Their links with other NGOs and government research organizations are often poor. In Lesotho, Sierra Leone, and even Honduras, the role of NGOs at times appears to dwarf the activities of the small research units within the government sector. Under these conditions especially, good overall policy direction and regulatory procedures are needed (see Box 2.6).

DEVELOPMENT PROJECTS AND AD HOC ACTIVITIES

In many countries short-term, donor-funded, rural-development projects that are not primarily focused on research nonetheless do a considerable amount of it, especially on-farm testing and introduction of new technologies. Research within donor-funded development projects may constitute a large share of the total research effort in a country. In some domains this share is out of proportion with that of the relevant national research institutions (Faye and Bingen 1989). This is particularly true for studies on natural resource management and for technology testing and adaptation.

Projects, however, are not formal organizations; they are fixed-term mechanisms used by donor organizations to implement an activity. Given the noninstitutionalized and often ad hoc character of the research performed by donor projects, research leaders in small countries seeking to maximize the benefits of such projects need to emphasize the coordination and information functions of their national system. It is vital, at the outset of negotiations with donors, to orient project activities around the objectives of the national agricultural research system as a whole. A survey of donor development projects in Lesotho identified over 50 projects carrying out agricultural research in the ministry (Okello and Namane 1992). Before this there had been little

---

**Box 2.6. Impact of Nongovernmental Organizations
on the National Research System in Sierra Leone**

The number of NGOs involved in technology introduction and development in agriculture and natural resources in Sierra Leone has increased dramatically in the past five years. The country's research organizations cannot hope to compete with the volume of advice and materials in crops, livestock, and natural resources that many NGOs currently provide. NGOs are particularly active in the south of the country, in restoring land that was hydraulically mined for rutile (titanium oxide), and in testing new crops and farming practices for resettled farmers. The NGOs have increased the national coverage of research and development and promoted community and farmer-based research strategies. There are three implications of this growing involvement of NGOs in agricultural research and development that policymakers and the managers of public research organizations must keep in mind.

First, there are increased regulatory responsibilities. In some cases, NGOs have brought germplasm into the country in diplomatic bags, bypassing official screening and quarantine procedures. Besides the fact that these materials receive little or no screening before being given to farmers, there is the danger of introducing new pests and diseases. Indeed, there is a strong suspicion that the sweet potato scab was introduced into Sierra Leone by an NGO.

Second, NGO involvement affects the resources of the core national research system. NGOs are a potential drain on its staff, as researchers and technicians are lured away by the better salaries and benefits offered by the NGOs. Wisely, the formal research organizations have chosen to establish complementarity with the work of NGOs rather than outright competition for staff and project funds.

Third, there are demands placed on the coordination and monitoring functions of the national system. With the pace of rural development accelerating, national institutions increasingly have important policy and linkage roles to perform. The revitalization of the National Agricultural Research Coordinating Council (NARCC) provides a forum for the coordination and monitoring of research and technology-transfer activities by projects and organizations outside the core national research system.

Source: Dahniya (1993).

---

attempt to compile a total picture of the scope of research in these externally managed projects. Some of the monitoring and coordination should fall to government organizations that bear a responsibility for research policy and coordination.

During project implementation, it is important to channel and store the information generated by a project so that it contributes to the total knowledge base of the national system. In this way, project outputs can be "institutionalized" and can in some measure increase the scale of research in small countries. However, this will not happen without planning, good coordination, and close links to policymakers.

Other potentially significant ad hoc research activities are those managed by farmer and producer organizations. The situation in Honduras, Mauritius, and Jamaica illustrates several instances where a group of producers have organized to contract research or to transfer technologies from outside. This is often the case in high-value nontraditional commodities in which producer associations may contract with scientists in countries where the technologies are in use in order to adapt and introduce them to local conditions. In Central America, producer groups were among the first to organize research on several nontraditional exports from melons to shrimps (Byrnes 1992; Rojas 1993; Zacarías 1992). The research organized and purchased by these producers was ad hoc and short term in nature. Their aim is to buy the research they need for use as quickly as possible, not to build research institutions. Here the ad hoc nature of research is a positive feature, provided the formal research system is aware of these activities. In risky and rapidly changing domains such as nontraditional high-value exports, ad hoc research can be a wise course of action.

DEALING WITH DIVERSE INSTITUTIONS

Data from small countries show that there is substantial diversity in the number and types of institutions involved in agricultural research and, because of the size constraint, most institutions are small. In many cases, however, consolidation into larger organizations does not seem to be an appropriate organizational strategy. Coordination at the policy level is often a more effective way to enhance the scale of research in a country. Conclusions from the ISNAR study suggest that organizational innovation and an emphasis on technology and information management, along with policy interactions, are as important as technology generation, per se, for building viable research systems in small countries.

THE CASE FOR COORDINATION

It is sometimes argued that small research systems require less coordination than large ones. However, given the considerable research capacity outside the government sector that is typical of small countries, governments must ensure that they are kept informed about what these other agencies are doing. And they must link and coordinate the activities of these institutions with those of the institutions they themselves support. Thus, in small countries, coordination becomes a critical function and responsibility of the national system. However, at present few countries have bodies or mechanisms that facilitate this coordination.

Coordination is essential for three reasons. It ensures that all participants in the national system orient their work around national development goals. Second, it allows policymakers to determine the comparative advantages of different institutions to conduct research within various domains, encouraging complementarity and avoiding duplication and competition. Third, the results of the research need to be collated at the national level to stimulate spillover between programs, institutions, and sectors.

RESEARCH COORDINATING BODIES

Case studies from the small-countries project show how increasing the capacity of publicly administered research institutions to provide policy direction and coordination enhances the performance of national research systems in small countries (Contreras 1992; Reid 1992). In some small countries, government research units have been reorganized to perform this policy and coordination role more efficiently. In Mauritius, for example, the research and technical services of the Ministry of Agriculture, Fisheries and Natural Resources have been restructured to create an Agricultural Policy Analysis Unit to complement the Directorate for Agricultural Research and Extension and the work of the Food and Agriculture Research Council. With three bodies in Mauritius charged with coordination of agricultural research, however, there is the risk that the three institutions may compete for the task of coordinating each other, resulting in less time available for research (Manrakhan 1992).

Sierra Leone has established an Agricultural Research Coordinating Council whose role is to orchestrate the country's various research components around a common policy. This policy integrates research on forestry, livestock, and staple food and export crops with research on land use and socioeconomics. It has helped ease the pressure on each institution to cover all the areas implied by its mandate (Dahniya 1992). Togo has recently created a National Directorate for Agricultural Research within the Ministry of Rural Development in an attempt to develop a coherent system out of a wide variety of institutional actors, ranging from donor-managed projects through parastatals to multicommodity national institutes (Aithnard and Gninofou 1992). The result in both countries is that policymakers and research managers are better able to plan and allocate research investments. Coordination at a higher level enables them to tap the various sources of funding, expertise, and knowledge scattered both within and outside the country. In contrast, amalgamating their diverse research efforts would produce a single institution whose scale could not have been supported from either private or public sources.

Research coordinating councils such as those in Sierra Leone, Mauritius, and the Gambia work under their respective ministries of agriculture and natural resources. In some countries, however, the council is interministerial, encompassing all science and technology activities relevant for agriculture, forestry, fisheries, and, sometimes, other spheres such as medicine and education. The Congo coordinates its five agricultural research institutions under a single Conseil National de la Recherche Scientifique et Technique (CNRST) within its Ministère de la Science et de la Technologie. Within the council, active coordination is assigned to a national directorate, the Direction Generale de la Recherche Scientifique et Technique (Onanga 1992). As research grapples increasingly with issues of natural resources and technology transfer, this broad approach to science and technology may prove more effective.

Some research coordinating councils serve as a clearinghouse for research funding. Donor and government funding can be channeled through councils to avoid competition for scarce funds. While this function may not always be welcomed by the institutes being coordinated, it is a useful way of focusing resources and activities on the priority issues, avoiding duplication, promoting shared programs, and identifying the comparative advantages of different institutions. Councils can also be an effective means for monitoring and channeling the activities and outputs of donor projects and NGOs (Dahniya 1993).

Lesotho's single government research division is hard-pressed to provide policy guidance and coordination for the many research and development activities that take place in donor projects throughout the country. As a result, little of the useful information and few of the technologies generated by these projects have been fully used. In other cases, technologies have been introduced that have neither promoted national development goals nor been integrated with other development efforts. For example, projects involving livestock development have introduced innovations that are at cross-purposes with projects on agroforestry and reforestation. One donor project is planting trees while, on the same mountain, another is introducing more goat varieties.

For very small systems, like Lesotho, Guinea-Bissau, Tonga, and Western Samoa, the research done in development projects may outweigh that of the research department. In Lesotho, where soil conservation is the top priority for research and development, the ministry's research division is dwarfed by the scope and range of research on conservation that exists within donor projects. Trying to cover what the projects cover is not an option, but obtaining and verifying their results is an important challenge that very small departments cannot ignore. The more ad hoc research there is, the greater the need for research institutions to emphasize information management and research coordination. Single research divisions or departments placed at low levels of the government structure may not be able to perform these essential policy and coordination functions.

CONSOLIDATION VERSUS COORDINATION

Most of the small countries in our sample suffered from an acute shortage of trained scientists. People with management experience were even scarcer. Dispersing this small core of scientists across several institutions multiplies the administrative costs of research and fragments the scientific community, making it difficult to build multidisciplinary teams and to achieve a critical mass of scientists to address specific issues. Because of this, it is commonly thought that a small country cannot afford to have more than one institution conducting agricultural research for national development goals. Consolidation, it is argued, can reduce costs as well as the bureaucratic distance between institutions. It can also promote a systems approach to research and enable a resource management perspective to be introduced.

There is truth to these arguments, but consolidation in fact appears advisable only under certain clearly defined conditions: when there is more than one research unit within the government (either in a single ministry or in separate ministries) and when fully fledged multidisciplinary commodity programs are simply not feasible or advisable, either because the country is too poor or because no single crop dominates agriculture to an extent that justifies major investment. Among the sample countries in the ISNAR study, Bhutan fulfills these conditions. It recently consolidated its previously separate government research units in crops, livestock, and forestry (Dorji, Gapasin, and Pradhan 1992).

Where there are several research institutions and units, each with different status, linked to specific subsectors of the agricultural economy for both funding and policy inputs, consolidation is usually not advisable for three main reasons.

First, consolidating private and parastatal research units under a single public structure can strain the fiscal capacity of the government. Many small countries are already overwhelmed by the costs of maintaining a modern nation-state. The additional burden imposed by adding more actors to the government research system may be too great.

Second, attempting to merge institutions whose cultures, management procedures, and salary structures are radically different risks causing severe administrative disruptions, as well as morale problems among staff whose status and conditions of service may suddenly deteriorate.

Finally, the value of contributions from nongovernmental agencies lies precisely in the ability of these agencies to tap and channel resources in ways not open to the government sector. Incorporation may stifle their freedom for innovation in areas beyond those covered by core institutions. These are the major arguments in favor of retaining institutional diversity and seeking other, more flexible means of promoting improved system integration and performance.

Another approach is to hand research currently done by the public sector over to the private sector because of the inherent flexibility and responsiveness to markets associated with the latter. This was the intention behind the establishment of private research foundations (Fernandez 1992; Wilson 1992). Privatization was a fashionable approach during the 1980s, when many small developing countries came under pressure to mimic the trend towards privatization so strongly prevalent in the developed world.

However, in most small developing countries, privatization is not advisable. The results of much of the research in staple and minor food crops, livestock systems, and natural resource management are not easily appropriable and would not be of interest to profit-making organizations. With their major concern for equity, public-sector research organizations are best placed to address the needs of the resource-poor farmers who make up the vast majority of producers in these countries.

Foreign research institutions located in small countries are a special case. For countries such as Togo, Congo, Vanuatu, and Chad, assuming the full burden of research that is currently done under the umbrella of the Centre de Coopération Internationale en Recherche Agronomique pour le Développement (CIRAD) is beyond national fiscal and human capabilities. Consolidation with a foreign research institution at the institutional level is not possible; however, improved coordination may be.

MANAGING COMPLEX FUNCTIONS

For the countries included in our study, organizational growth is not an option. But what does it mean for a research system to remain small? It means that there will be fewer experiments and studies, and fewer staff. Also, as scale becomes smaller, the functional complexity of an organization's role and tasks is greater. In small countries, it is difficult to separate traditional research functions such as experimentation from other functions like technology transfer, policy advice, coordination, and regulation. In other words, researchers and managers in small-scale organizations will have more complex functions with more horizontal responsibilities than those in large organizations.

HIGH-QUALITY STAFF COUNTERBALANCE SCALE LIMITATIONS

If complex functions, quality, and greater attention to management and policy issues are expected from researchers in a small country, these countries will require further investment in improving the quality of their staff. Data on the qualifications of agricultural researchers in 31 small countries show that most researchers have only a first degree, a little more than a third have Masters degrees, and only 8.5% have a PhD. While we cannot argue for more researchers, this ratio of 1 PhD to 7 MSc to 10 BSc degree holders supports the argument that researchers in small countries need more training. They must be as well qualified as possible if they are to function effectively in a small NARS: they have to be able to interact comfortably at the national policy level as well as with the global community of agricultural science.

The ratio of researchers holding a PhD or MSc to those with only a BSc needs to be more evenly distributed in small countries through greater investments in higher-level training. This is necessary even if the sole objective is to increase "borrowing" capacity. Improvements in the quality of human capital increase a country's ability to borrow technology and make better use of technological spillovers. For small countries, investing in human capital is more important than investing in research facilities.

CONCLUSION: SMALL SCALE CAN BE AN ASSET

Small scale, although usually seen to be a constraint, is not all bad news. It also offers opportunities for innovation in the quality of the research produced and in the advice that is given. Quality and scope are areas where small countries can diversify and emphasize their comparative advantages. Small-country research organizations can look to current trends in international business where excellence is increasingly seen to be linked to "smallness" in both scale and attitudes.

Increasingly, large companies want to act like small companies and take on the advantages of product focus and quality that small scale can support. According to Peters (1992: 560), "almost all big firms are working overtime to try and act like small firms". They see smallness as a desirable characteristic. In stable economic conditions, organizations can plan and compete on scale; during periods of instability, they must stay small and flexible, and compete by living off their wits (Peters 1992). Management and production mechanisms like networking, "outsourcing", and "spill-ins", to use the popular neologisms of the business world, make it possible to be small and use networking to achieve the scale to perform big tasks.

Rather than growing in size, successful small firms are innovative in their man-agement strategies and they tap new sources of products and information, while moving closer to their clients. There are obvious parallels with the strategies that small-country NARS can adopt. Since their research organizations can never be big, they have an opportunity to be innovative in their policy and management strategies. They can adopt strategies that build upon, and are derived from, their advantages as small organizations, rather than trying to emulate inappropriate strategies from larger countries that do not share their advantages.

In this chapter we have highlighted the need to accept the scale constraint and to explore the possibilities of using small size as an advantage by being more flexible, sharing responsibilities, and seeking organizational and strategic innovation. In Chapter 3 we illustrate in more detail how the portfolio approach is used to develop workable research scope, given the constraints of scale. It emphasizes the fact that small-scale institutions can address a broad scope of research through a broader range of functions and strategic and managerial innovation.

Notes

1. Four countries (Equatorial Guinea, the Maldives, Nauru, and Tuvalu) had no identifiable agricultural research institutions.
2. The distinction between core and supporting institutions is important: a core institution has a mandate and national character that place it squarely within the national research system; supporting institutions, such as NGOs or producers' associations, are not specifi-cally mandated to conduct agricultural research but often do so as part of their activities or for specific purposes.
3. We call it *core* because of its primary mandate for agricultural research and its central role in a particular area of research.
4. The average number of researchers for public-sector research institutions is about 20.

3 The National Research Portfolio: Establishing a Manageable Scope

INTRODUCTION: THE INCREASING DEMAND FOR RESEARCH

Agriculture and natural resources are the largest sectors in most small, low-income countries. In Guinea-Bissau, for example, 80% of the population depends upon agriculture for their livelihood, and even in countries where agriculture is not the main occupation, agricultural products are important sources of foreign exchange. Countries like Jamaica and Trinidad and Tobago, which were primarily agricultural in the 1960s and became dependent upon tourism and mineral exports in the 1970s and 1980s, have renewed their interest in agriculture and renewable natural resources.

The policy environment for agricultural development is marked by growing competition for export markets, deepening problems of rural poverty, and stagnating or declining production. There are increasing demands on agricultural research to produce new technologies to raise productivity and to identify new options for and sustainable uses of the natural resource base. And these demands are stretching the capacity of research organizations in small countries. Many research leaders are actually under increasing pressure from policymakers and producers to take on more research.

CAN SMALL COUNTRIES NARROW THE SCOPE OF THEIR RESEARCH?

Reviews of research scope in many developing countries conclude that national research systems attempt to cover too many areas, commodities, and agroecological zones. In small countries, we find many small institutions trying to cover the same range of topics and commodities as much larger institutions in countries with far greater resources. Since research scale is small, conventional wisdom and "expert" advice recommend that they must narrow and consolidate their scope to cover fewer areas.

Like much well-meaning advice, this seems very sensible. However, when a research leader tries to put it into practice, it is often far from easy. This chapter considers how small countries can narrow their scope to match their scale, whether they can afford not to cover certain domains of research, whether the research functions and the intensity of the research effort are the same for all research domains, and how institutions outside the NARS can be mobilized to cover areas beyond the capability of national institutions.

CAN SMALL COUNTRIES USE EXTERNAL TECHNOLOGIES AND INFORMATION SOURCES?

Small countries are often advised not to do research but to borrow technology developed elsewhere. Certainly, small countries have a lot to gain from a national research strategy that emphasizes scanning, selecting, and transferring technologies from abroad. However, suggestions that small countries only have to "beg, borrow, or steal" external technologies ignore three key points.

First, selecting technologies appropriate to local needs and conditions is complex. Screening, borrowing, and adapting scientific knowledge and technology require essentially the same capacity as technology development (Ruttan 1989). The scale of this capacity is usually underestimated, and in many small countries it does not exist. It is ironic that while small countries can benefit most from the spillovers of technologies generated elsewhere, using these same technologies is more difficult for small-country NARS than for their larger competitors. "Smallness is a greater barrier to the ability to benefit from the effects of international technology spillovers in agriculture than in other sectors" (Milner and Westaway 1993: 211).

Second, managing the scanning and linkage functions in NARS is particularly difficult for small research organizations. Even collaborative research networks intended to strengthen local efforts tend to either bypass or overload weaker research systems (Ruttan 1989; Houssou 1992; Gakale 1992). This is discussed in greater detail in Chapter 10.

Finally, it is clear that technology generation, selection, and adaptation require staff with similar levels of training and experience. Highly trained scientists will be needed in even the smallest systems. In very small research organizations where there are only one or two researchers to cover key areas, their qualifications need to be higher than the norm in order to cope with the multiple functions that research is expected to perform. Central among these functions are scanning world science and technology to select technologies for local adaptation, as well as providing advice to policymakers on technological and resource management.

Researchers in small countries often suffer from a sense of professional isolation; they have few contacts with other scientists and little opportunity to benefit from the advice and assistance of experienced senior colleagues. Operating independently or in very small teams requires staff to have broad, high-level scientific training, and not just in science. In a small system, each researcher is required to operate at a higher level within the national policy environment, as well as serving as a link with the global community of agricultural science. These tasks involve complex decision making and acceptance of major responsibilities, and they call for highly trained researchers who can manage a wide range of linkages, interact at high levels (often representing their country at international fora), and yet produce practical research results.

As discussed in Chapter 2, the ratio of graduate to undergraduate degree-holders in small countries needs to be more evenly distributed, and this can be accomplished through greater investments in higher-level training. This is essential even if the objective is to increase borrowing capacity. Improvements in the quality of human capital increase a country's ability to borrow technology and to make better use of technological spillovers. For small countries, investing in human capital is more important than investing in research facilities.

If "technology borrowing" is to be a feasible strategy, more attention is needed on how local research capacity can be institutionalized, allocated, and oriented to benefit from technologies generated elsewhere. This reorientation of NARS to make the most of external knowledge and technology lies at the heart of our *portfolio approach*.[1]

USING THE PORTFOLIO APPROACH TO MANAGE NATIONAL RESEARCH SCOPE

The portfolio approach has four elements: the first is to unify policy and coordination, second is to identify the scale of research, third is to set the scope of research, and fourth is to assign institutional responsibilities. Chapter 2 discussed the first two steps in the approach; this chapter focuses on the third and fourth steps (see Figure 3.1): setting scope and then assigning institutional responsibilities for different research domains and functions.

Setting research scope within a portfolio involves two activities. First, the potential scope of research is segmented into seven domains. Second, the research functions that institutions can be expected to perform are identified.

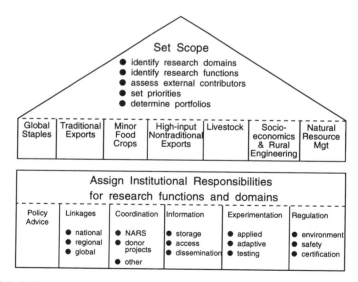

Figure 3.1. Setting the scope in a research portfolio

The final step in the portfolio approach is to assign responsibilities for specific research domains and functions to institutions, according to their mandates and capacity. Research planners in small countries need to pay more attention to the problems of how to organize and manage local research capacity so that it can benefit from the spillover of useful technologies generated in other countries (Box 3.1). This reorientation of national systems to enable them to make the most of external knowledge and technology lies at the heart of our rationale for applying a portfolio approach to the management of research.

Box 3.1. Coconut Research in the Solomon Islands: A Problem of Scope

Many small countries depend on a few traditional export crops for a large share of their export earnings. Given the strategic importance of these crops for generating foreign exchange, research is needed to improve and even merely to maintain their productivity. Yet no matter how large a proportion of total agricultural production the crop accounts for, its absolute value may not support a comprehensive research effort. Furthermore, most small countries simply do not have enough scientists to devote to major research programs in single commodities.

Coconut research in the Solomon Islands provides a good illustration. The sector needs (a) the results of applied research, providing new technologies or information of practical relevance to growers and (b) the results of adaptive research, taking the products of either local or overseas research and evaluating their applicability to local conditions. These research needs are not proportional to the size of the coconut industry: the Solomon Islands probably has as many research needs as the Philippines, though the Philippines produces as many coconuts in a week as the Solomon Islands does in a year.

The scope of the research outputs needed by the Solomon Islands is therefore as broad as it is in some very large countries, but the scale of its resources is not sufficient to do all that is needed. The portfolio approach allows the country to identify the sources of appropriate coconut technology and the linkage mechanisms required to obtain it. In this way it can focus its own research on key topics not covered by others. It can also become a more efficient scanner of technologies produced elsewhere. The portfolio approach also outlines the range of functions required of the research system when technology generation is not feasible, and it identifies a set of organizational options to perform those functions.

Source: Ilala (1989).

SETTING THE SCOPE OF RESEARCH

IDENTIFYING DOMAINS IN THE RESEARCH PORTFOLIO

Our studies have shown that the potential scope of research can be broken down into seven domains (Table 3.1), which cover various types of research: some are focused on commodities and production techniques, while others study systems and include information-gathering on resources or markets. Each domain in the portfolio implies a distinctive institutional and management strategy. The typology is particularly useful in identifying sources of research information (Ballantyne 1991).

The categorization serves as a guide to investments, actions, and partnerships for the organizations that comprise the national system. The typology developed for classifying research activities and topics into domains is based on the following criteria:

- the intensity of the demand for research within the domain;
- the policy environment for technology development within a domain;
- the existing network of technology sources, availability, and flows;
- the range of partners and organizations that can contribute to research.

Global staples are food crops such as grains, tubers, and grain legumes that are of global importance and which provide the bulk of dietary calories for rural and urban dwellers alike. National food security is a central concern of public policy in this area. All public research institutions in NARS have their major focus here. Global staples are also the focus of research work by international centers and, in some cases (such as maize and soybeans), the multinational private sector.

Internationally, research on these commodities is well supported and information is widely available. This makes it possible for small government-based institutions to work in the area of global staples despite limited national resources. Small countries need to carefully select the work that they do in collaboration with international centers, and to welcome the private sector. The strategy is to access results from outside in order to screen and adapt them for dissemination within the country.

Traditional exports include crops such as sugar, bananas, cotton, rubber, or cocoa upon which many small countries depend for a major share of their export revenues. Much of the research on these crops is done in private companies, parastatal commodity boards, or the Centre de Coopération Internationale en Recherche Agronomique pour le Développement (CIRAD). Access to technology is available but within a more restricted sphere, where producer groups can be an important link. In general, a small country will be continuously challenged to protect its industry from natural disasters and to lower costs to meet competition from other producers or new entrants to the market. Hence, research on these commodities tends to be institutionalized.

Minor and traditional food crops are important to the local food-producing sector but are "orphans" in international trade. Hence, they are often not the focus of

Table 3.1. Categories of Research Domains

Global staples	Traditional export crops	Minor food crops	High-input, nontraditional export crops	Livestock	Socioeconomics and rural engineering	Natural resource management
Beans	Bananas	Apples	Asparagus	*Small Ruminants:*	Farm management	*Fisheries*
Cassava	Cashew nuts	Barley	Broccoli	Goats	Farm structures	
Cowpeas	Cinnamon	Breadfruit	Brussels sprouts	Sheep	Farming systems research	*Forestry*
Groundnuts	Cloves	Broad & mung beans	Cardamom			
Maize	Cocoa	Cabbage	Citrus (limes, grapefruit)	*Large Animals:*	Marketing research	Agroforestry
Potatoes	Coconuts	Carrots	Flowers/ornamentals	Cattle	Postharvest and storage	Genetic resources
Pulses	Coffee	Castor beans	Fruits	Horses	Machinery and tools	Plant pest & disease management
Rice	Cotton	Date palms	Ginger	Camels	Irrigation	
Sorghum	Oil palm	Figs	Grapes	Donkeys	Rural engineering	
Soya	Rubber	Fruit (local use)	High-value vegetables		Agroprocessing	*Land Use and*
Wheat	Sisal	Garlic	Jojoba	*Small Stock:*	Agroindustries	*Water Mgmt:*
	Sugar	Lentils	Kava	Chickens	Agricultural wastes	Soil (fertility,
	Tea	Millet (*Eleusine, Digitaria*)	Litchi	Ducks		erosion,
	Tobacco	Mustard (seed)	Mangoes	Turkeys		conservation)
		Oats	Melons	Swine		Water resource management
		Okra	Palm hearts			Range & pasture
		Onions	Papaya	Animal health		
		Pandanus	Passion-fruit	Feeds and nutrition		
		Peas (garden-)	Peaches	Animal breeding		
		Pears	Pineapples	Wildlife management		
		Peppers	Plums			
		Pigeon peas	Pyrethrum	*Aquaculture*		
		Plantain	Quinquina			
		Radishes	Ramie (textile fiber)			
		Safflower (oilseed)	Soursop			
		Sesame	Spices			
		Sunflowers	Starfruit			
		Sweet potatoes	Strawberries			
		Swiss chard	Sunflowers			
		Taro (*Xanthosoma, Colocasia*)	Vanilla			
		Tomatoes	Ylang-ylang			
		Triticale				
		Turnips				
		Vegetables (local use)				
		Yams (*Dioscorea*)				

Source: Eyzaguirre (1991).

large-scale research at the global level. Information and technologies on taro, plantains, sweet potatoes, and grains and legumes such as fonio and bambara groundnut are not easy for developing countries to get,[2] nor are they well supported. Small-country NARS may have to do applied research to compensate for the dearth of appropriate technology globally. Fortunately, some sources of assistance, such as the Asian Vegetable Research and Development Centre (AVRDC), are now in a position to provide low-cost access to germplasm. However, traditional foods peculiar to a given country (e.g., fonio, taro) will continue to require local research.

High-value nontraditional exports are important in the diversification strategies of many small countries. A large part of the research on high-value nontraditional export crops is private and ad hoc and must be contracted for or purchased. Again, producers and private companies may be set up to obtain this type of research. Consultancies and transfer of existing technologies are the main mechanisms used to meet production needs. For these commodities, postharvest technologies, transport, handling, and marketing are the key areas for research, which should also be done in conjunction with producers. Chapter 6 describes how research on these high-value commodities is best carried out in conjunction with a product development cycle.

Livestock research on all aspects of animal production and health is important for all countries. With some notable exceptions, such as Botswana, Namibia, Paraguay, and Mongolia, however, most small countries do not make significant investments in livestock research. There are two basic types of research. Research on poultry, swine, and industrial dairying to serve urban markets and their associated processing industries is largely in the hands of private-sector companies. Research for less-intensive production systems, where livestock such as small ruminants, dual-purpose cattle, and camels are raised for both social and economic reasons, tends to be the responsibility of the public sector.

While not all countries can justify animal breeding programs, most require more research work to monitor the state of national herds and livestock production systems, including interactions with crops, water, and vegetation. Integration of livestock research with research on natural resources may allow considerable economies of scope to develop and increase the attention given to livestock in general. This kind of research to assess grazing conditions, stocking rates, and disease prevalence depends heavily on information and tends to fall within the regulatory and advisory functions of small NARS. Government research agencies are the dominant actors here.

Socioeconomics and rural engineering includes socioeconomic analysis of farmer choices, production constraints, postharvest technologies, farm structures and machinery, market research, and policy. A small-country research system has a comparative advantage in understanding the local social and economic constraints facing its own farmers. This gives it an advantage, as well, in influencing the policy environment and in providing advice for intelligent borrowing of technology. However, this type of research is underrepresented in most of the NARS in our study. Research in this domain is country-specific but it employs widely applicable methodologies. Because it is closely linked to policy considerations, the aim of many socioeconomic studies is to produce information for policymakers to use in setting agricultural policies. Also, implementing policies on diversification requires more postharvest and marketing research. Much of the research capacity in these areas is found outside the crop and livestock research institutions; it is in universities and in planning and policy units within government ministries.

Research on natural resource management is aimed at managing an existing resource, often location-specific. At the applied level, it involves both local and nongovernmental organizations. It also requires strategic and long-term research at a scale that may require transnational efforts. Global, regional, national, and local information has to be assembled and made available to policymakers so that they can make informed decisions on the most productive and sustainable use of the country's natural resources. Some of the research is done at the local level in conjunction with farmers and other resource users; NGOs are active at this level. More policy-oriented research, which is aimed at monitoring and measuring environmental processes in agriculture and which takes intersectoral issues into account, is more likely to be done within universities or by specialized environmental institutes. Some factor-based research and measurement is done regionally. Establishing the division of labor to undertake and collate these studies is a key task of public agencies that can link to policy and provide a long-term perspective. Chapter 5 presents a more in-depth discussion of the impact of natural resource management on small-country research portfolios.

OVERVIEW OF DOMAINS AND THE SCOPE OF RESEARCH

A survey of the research scope of 45 core NARS in 23 small countries for which we were able to obtain complete experiment reports revealed a concentration on global staples and traditional export crops.[3] This is as expected, considering the emphasis and support given to these research areas by the development community. The diagram in Figure 3.2 indicates a tendency for small-country NARS to mimic the concentrations of research activities in larger NARS and international centers. An alternative approach for these countries might be to shape the national scope of research in ways that establish complementarity with external efforts, by building a national capacity to select research concentrations based on their suitability to local conditions. This might mean that the focus on global staples and traditional exports could be reduced somewhat to reallocate those resources to more location-specific research themes in socioeconomics and natural resource management. This would also enable countries to be more effective in their scanning and selection of available technologies and information in those domains where technology, information, and support are abundant. This might be one way to reduce the amount of duplication in technology generation and experimentation that is sometimes found in small-country NARS.

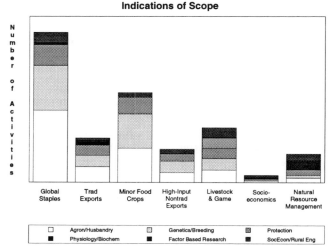

Figure 3.2. Schematic diagram showing disciplinary focus across research domains

DETERMINING THE RESEARCH FUNCTIONS TO BE PERFORMED

The second step in applying the research portfolio approach is to identify the NARS functions that need to be performed so that the demands in the various segments of the portfolio can be met. In the past, national research systems were conceived as a set of organizations conducting experiments, whose primary goal was to generate new technologies. In small countries, the functions of national systems have to be much broader. Responsibility for the different functions required in each domain is allocated to different institutions based on their capacity and comparative advantage.

Policy advice. Government research services in the smallest countries (with fewer than 25 researchers) have a crucial role to play in providing advice for policy formulation (see Box 3.2). Reviews of agricultural-sector strategies in the South Pacific, for example, have noted the lack of capacity to analyze policy options, formulate informed agricultural policies, and represent the interests of farmers and resource users in the policy-making levels of government (World Bank 1990, 1991). Policymakers are often faced with urgent decisions that cannot wait for NARS to design programs and conduct experiments in order to provide an informed answer.[4] Policy advice from NARS must be based on the results of its research programs and from a review of current information.

Managing linkages. Many of the resources needed for conducting research—information, money, and expertise—exist outside small research institutions, and often outside the country. Linkages with producers and policymakers within the country are also essential in order to focus scarce resources on priority problems. Establishing and managing such linkages is an essential function within the NARS. It must be

---

Box 3.2. Agricultural Research for Policy Advice

In Trinidad and Tobago, science and technology are of crucial importance to agricultural development. However, the country cannot afford to sustain a single institution within the state, private, or parastatal sector that can address all the problems. In a market-oriented economy that emphasizes productive flexibility, policymakers need to receive up-to-date technical advice on the options for agriculture and the use of natural resources. This demand on research from policymakers is primarily for information for decision making with regard to the following:

- the choice of key commodities requiring priority attention, whether for domestic or international markets;
- the determination of the ways and means to effect significant improvements in agricultural productivity while conserving the natural resource base.

Source: Rudder (1992).

---

noted, however, that some mechanisms for establishing linkages, such as networks and regional research collaboration, impose greater demands than others on the abilities of small-country NARS.

Coordination. It is likely that no single research institution in a small country will be able to address all of the priority research topics within the national scope; several institutions will have to be mobilized. The fact that there may be several external agencies, national institutions, and NGOs that conduct research or introduce technology on a short-term or ad hoc basis also creates a need for coordination. The way this core function is performed varies from country to country. But two key conditions for success are close linkages to the policy level and a forum for decision making and participation by the various components of the NARS and the external contributors or partners.

Managing information. Managing scientific information is a key function of NARS in small countries, and gaining access to relevant sources of information is the rationale for establishing most external linkages. Scanning, selecting relevant external sources, and directing useful information to producers and policymakers is a major way for research systems to perform their tasks, without necessarily generating new technologies or conducting extensive experimentation. More information is available today than ever before, with the number of sources increasing daily. The abundance and relevance of these sources varies across the national scope of research, so NARS information managers need to work in close partnership with managers of research programs to maximize the benefits of technology spillovers from countries with similar problems but more resources.

Experimentation. This is the traditional function of a NARS—one that remains essential but which may no longer be the primary source of information and technologies. Small-country NARS should focus on areas where there is little or no information or technologies available from outside; this is where they need to concentrate their experimentation, technology generation, and adaptation. Some-times, a small country can move into upstream applied research. However, unlike large, wealthy countries that can afford to begin with basic research and then to move into applied and adaptive research, the small country begins with testing and moves into adaptive and even applied research as conditions and capacity warrant. Their approach must be from the bottom up.

Regulation. In many small countries, much of the agricultural research and development takes place outside public-sector organizations with national research mandates. It may be unrealistic for these research organizations to assume responsi-bility for the activities of external agencies, NGOs, and development projects. However, it is important for the public research organization to be responsible for regulating technology introduction in the country. This can avoid some of the misdirected and often harmful effects of inappropriate introduction and use of technologies. Quarantine, seed production, and guidelines on agricultural inputs are some of the regulatory areas where NARS should contribute.

ALLOCATING INSTITUTIONAL RESPONSIBILITIES IN RELATION TO DOMAINS AND FUNCTIONS

This strategic framework encourages research institutions to set their priorities based not only on what they can do, but also on what others are doing or are able to do. Policymakers may identify a particular agricultural constraint as the most important problem for research to solve, yet allocate few public resources to it. Often another organization is better able to do the work, or, information and technology from past research elsewhere is readily available. In that case, it may not require major in-country research efforts. If it is a long-term information-based activity such as land-use monitoring or socioeconomic studies, then a university may be a better location for the research; if it is a high-risk, high-return technology initiative, then it may be best left to private-sector producer groups.

The way in which the system is structured will reflect these choices: the technical conditions under which the research is carried out and the availability of partners. The objectives of this approach are (a) to identify the comparative advantages of different institutions to work in the various domains of the national research portfolio, (b) to guide decisions about the allocation of responsibility for research, and (c) to promote complementarity between the actors in the system.

Our approach is to bring together the research domains, the functions, and the institutions in a decision-making process that allocates responsibilities to the various actors (see Figure 3.3). The decision to allocate responsibilities for research depends on the relative strengths of these factors: policy pressures or support, market condi-tions, the scale of research needed, and institutional complementarity.

High-value Nontraditional Exports	Socioeconomics & Rural Engineering	Traditional Exports	Global Staples	Minor Food Crops	Livestock	Natural Resource Mgmt

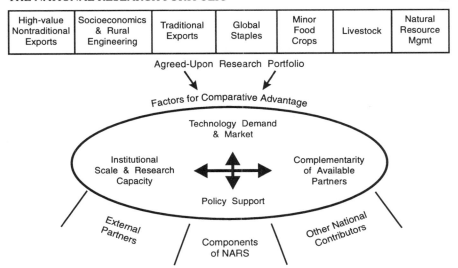

Figure 3.3. Factors in determining the comparative advantage of institutions

Policy support. Agricultural research is both in the knowledge generation business and the development business. Policies established by the government affect the way in which the system will gravitate both directly (through resource allocations) and indirectly (through demand for technology).

Market demand. Export crops, whether traditional or nontraditional, are closely integrated with development and marketing. While research on these crops must be located in the appropriate agroecological region, it is often separated institutionally from the rest of the country's agricultural research. This reflects its orientation towards the market and the policy environment.

Institutional scale and research capacity. The decision to work on certain problems imposes a scale of operation that determines the location of research. For example, certain types of research (e.g., farming systems research) are necessarily decentralized, while new research techniques, such as biotechnology, are oriented towards universities and sources of knowledge. Biotechnology and advances in information technology and communications are changing the way research is conducted and the location of many types of research. Simulation models and biotechnology are shortening research lags. Improved access to technologies developed elsewhere is changing the scale of investment needed to apply such technologies. All this is forcing structural change.

Institutional complementarity and partners. Small systems are more susceptible to biases introduced by external agencies, both technical and financial; a small program of collaboration can exert a major influence on the priorities of a small system. The availability of external partners may also be an important reason for a small country to get involved in research, as is the case with global staples. Where partners are not available, the levels of national investment will be higher and a small NARS will have to consider very carefully the functions it will perform in an area where it will essentially be alone.

These factors all represent "pulls" on the structure of a national research system. Different types of research require different structures. We may expect small systems to appear fragmented as they try to ensure that suitable structures are in place to support all their research priorities. Indeed, some small systems come under such extreme stress as a result of these forces that they risk losing all coherence and can hardly be said to constitute a single system at all.

These factors place an added burden on the task of research planning, which can become a highly volatile process as a result. As the objectives of the system change, so will pressures arise for structures to change as well. If there is not enough consensus for structural change or if such change would risk disrupting existing programs that are currently performing well, new linkage mechanisms and joint programs may be needed to ensure that new research topics can be addressed.

In assigning responsibilities for various NARS functions, the portfolio serves as a template upon which priorities and decisions to invest in research are made. It involves choices not only among commodity and research areas, but, within those areas, choices among a limited number of research objectives.

The research production function is likely to be different across commodity and research areas. If committed to certain research areas, any country (whether large or small) will require the same minimum critical mass and have to perform the same research functions. For example, if a country is an important producer of vanilla (such as the Comoros) or its population attaches great importance to a local variety of sticky rice (as in Laos), the country will have to maintain a full research program on that product. In other areas, a small system may adopt a different research strategy: where a larger system would decide to maintain a full breeding program, the small country could take essentially a borrowing strategy to address production priorities.

Table 3.2 shows a research strategy for a small research system, planned using a portfolio approach. For each of the seven domains, it presents a set of objectives and identifies the target group to benefit from the research. The approach then suggests a research strategy based on the availability of information and technology, names the potential suppliers of research, and outlines the functions and skills that the research system needs in order to perform.

The decision-making model is not a checklist where all factors need to be present in equal measure or in a given proportion. Rather, it serves as a guide to how the various factors act as pulls on a system. For example, external partners can compensate for the lack of institutional scale; policy can play an important role in creating demand for research in a domain where the market forces are weak, as in the case of subsistence crops or natural resource issues. Some countries have built up institutional scale by concentrating on single commodities over a long period of time, and

Table 3.2. Applying the Portfolio Approach

Research Portfolio	Objective	Target Group for Research	Research Strategy	Sources	Functions & Required Skills
Global staples	maintenance of productivity, local germplasm conservation, farming systems	small farmers	screening, dissemination, farming systems	IARCs, other NARS, networks, donor countries	information management, network management, adaptive & on-farm research
Traditional exports	product quality, productivity, maintenance, crop protection, cost reduction	plantation sector, smallholder outgrowers	breeding for quality, yield, crop protection, cost reduction	agribusiness, parastatals, commodity institutes, commodity associations	breeding, postharvest technologies, plant protection
Minor food crops	maintenance, conservation local germplasm, new products	small farmers, traditional producers	screen imported varieties	larger NARS	information management, crop protection, local farming systems knowledge
High-input nontraditional exports	product quality, agronomic practices, timing for market windows, new products	specialized farmers	import vertically integrated operations, focus on post-harvest & marketing research	agribusiness, universities	market intelligence, management of public-/private-sector links
Livestock a. high input b. traditional	a. production for specialized urban markets b. improved races, animal health	a. commercial farmers linked to markets b. OFR on dual-purpose: animals & mixed farming systems	a. animal health, regulation, postharvest operations b. animal health, on-farm research, nutrition	a. livestock industry, government institutes b. government institutes, NGOs, development projects	a. market research, veterinary research b. on-farm systems research, nutrition, socioeconomic research
Natural resource management	a. appropriate national environmental policy & regulation of land & water use b. sustainability research for fragile areas	a. policymakers b. borrow strategic research, conduct farm-level research	a. access environmental research policies of others b. access strategic research, conduct "precision research" c. land-use surveys, recommend regulations	a. universities, public-sector research & development agencies b. IARCs, NGOs, public-sector research institutes c. universities, development agencies, regional & global environmental agencies, e.g., FAO, UNEP	a. information management, policy analysis b. access information, regulatory strategies; OFR c. information management, land-use planning, geographic information systems (GIS)
Socioeconomics and rural engineering	a. appropriate national R&D policies b. knowledge of resource base, constraints, production systems c. post-harvest handling, storage, marketing	a. policymakers b. planners, on-farm research scientists c. producers	a. national strategic planning b. on-farm research c. screening & testing technologies	a. universities, government research & planning bodies b. public-sector research, NGOs, develop't agencies c. agribusiness, producer organizations, NGOs, development agencies	a. macroeconomic & policy analysis b. farming systems economics, anthropology

Source: Adapted from Elliott (1992).

this institutional scale can serve as a pull to engage in types of research where support from policy and external partners is weak. The model is used to indicate how the four factors that influence choices and options interact to determine the institutional comparative advantage.

Chapter 4 presents case study examples of how different factors influence the national research portfolios in small countries under different conditions. It provides more concrete illustrations using the portfolio approach.

Notes

1. The portfolio-management approach used here elaborates on the portfolio analysis that is used to assess the business performance of firms. We link our portfolio-management approach to the analysis of markets (domains, competitors, and available partners) and products (types of technology or research) that are needed (see Davies 1991: 141–142).
2. With the creation of the Interational Network for the Improvement of Banana and Plantain (INIBAP) and the expansion of Centro Internacional de la Papa's (CIP) work on sweet potatoes, information and technologies on these crops are likely to become more readily available to small developing countries.
3. These data were assembled from NARS's annual research reports and reports to the Current Agricultural Research Information System (CARIS) of FAO, which codes activities according to specific classes. The coded activities were then grouped under our seven research domains. It is important to note that the illustration should not be seen as a measure of research intensity and resource allocation, or of the importance given to a particular domain. The codes merely reflect the objectives of the research activity. The number refers to the number of separate activities within a group; it does not refer to the amount of resources or intensity or duration of the activity.
4. Research is a valuable policy tool. It serves development policy in two ways: one is by generating new technological inputs that can make agriculture more productive and efficient in its use of resources; the other is by providing policymakers with information on agricultural potential and the optimal use of resources. Some of these issues are discussed in more depth in Schuh and Norton (1992).

4 National Research Portfolios: Case-Study Findings

THE COUNTRY CASE-STUDY APPROACH

National research leaders from seven countries (Togo, Sierra Leone, Mauritius, Lesotho, Fiji, Jamaica, and Honduras) prepared country case studies on the composition and structure of their national research systems. This provided a range of cases that made it possible to analyze four areas in research policy and organization:

- institutional complexity and diversity in the composition of national research systems and how this affects national research capacity;
- emerging issues, such as natural resource management or diversification, and how national systems address the pressures to expand the national scope of research that these issues create;
- the size and structure of the NARS and how it affects the research portfolio and the choice of functions that can be performed;
- participation by small-country NARS in regional and international research partnerships and how the opportunities and management problems associated with these transnational partnerships compare across countries and regions (Box 4.1).

COMPARING SCALE, INSTITUTIONAL DIVERSITY, AND FUNCTIONS OF SMALL-COUNTRY NATIONAL AGRICULTURAL RESEARCH SYSTEMS

Standard measures of research intensity or scale, based on ratios of research investments as a percentage of agricultural GDP, per unit of agricultural land, or per unit labor, all show that small countries have higher agricultural research-intensity ratios than larger countries (Pardey, Roseboom, and Anderson 1991: 295). However, below a certain scale, country comparisons with indicator ratios may be misleading, given the actual amount of research that small countries are able to mobilize and sustain. In the case studies, we used available data on numbers of scientists and resources to produce an overview of the institutional composition of NARS.[1] Given that dramatic increases in the number of scientists and facilities and funds are difficult to sustain in small countries, the studies focused on how research is institutionalized and coordinated in a national system. This is as important an indicator of stability and capacity to produce relevant outputs as the numbers of scientists and expenditures.

Box 4.1. Why We Chose the Case-Study Method

The case-study method is a central technique in anthropology and medicine and is perhaps best known by the central role it plays in management and organizational research and training. It is a useful method for considering complex systems with a wide range of factors. In order to use the method successfully, premises and units of analysis need to be carefully defined in advance. This was done at the first small-country NARS workshop in February 1990, where research leaders and advisors met to agree on the methodology (see Eyzaguirre 1991). The selected country case studies were designed to produce valid and verifiable findings about the processes and environments of research systems (Yin 1984: 97).

Case studies are particularly useful in assessing the degree to which organizations and systems achieve their goals. The growing concern with "what is and what ought to be" has led to increased use of the case-study method. It is extremely valuable for the holistic study of complex systems and the process of change.

The case-study approach is also used to describe and analyze phenomena from the point of view of those involved. The ISNAR small-country case studies were produced by research leaders from small developing countries and are designed to reflect their perceptions of the opportunities and constraints they face.

Each case study drew generalizations that are applicable to similar situations, as evidenced by the "shock of recognition" in those who read the cases. Case studies are also tools for theory building: using inductive, pragmatic, and concrete strategies based on holistic data gathering. The strategic approach presented here, which proposes a research portfolio approach to build upon institutional diversity and broad NARS functions, was the result of the case-study findings.

Source: Adapted from Coolen, Beal, and Moran (1984).

The seven country case studies of organizational structure and the four companion country studies on scientific information covered the full range of institutional diversity. Honduras was the most diverse, with research institutions in the public, private, parastatal, and university sectors (see Table 4.1). At the other extreme, Lesotho had only one very small research division placed in the lower levels of the ministry of agriculture.[2] A comparison of the approaches in our selected cases illustrates the range of options and strategies to achieve stability, flexibility, and institutional diversity that are the hallmarks of effective research systems in small countries. Honduras presented the clearest example of this.

The case studies also compare the range and mix of functions that NARS perform. Mauritius has invested the bulk of its resources in generating technology and has concentrated them in a single domain. In contrast, Sierra Leone gives greater emphasis to functions such as the management of linkages with farmers and sources of technology. The Seychelles undertakes little or no research and focuses primarily on the access and management of scientific information. An important function that was underemphasized in all the case studies was the contribution the NARS can make to policy and decision making. Even where technology generation is not the central task, NARS are

Table 4.1. Institutional Composition of NARS in the Case Studies, 1990–1991

Country	Government Agricultural Research Organizations	Parastatals	Foundations	Private Firms	University-Based Research Institutions
Honduras	2	1	1	2	2
Jamaica*	1	4	1	–	1
Togo	–	–	–	–	–
Sierra Leone	2	–	–	–	1
Fiji*	3	1	–	–	–
Lesotho	1	–	–	–	–
Mauritius	2	1	–	1	1

* These countries have regional organizations executing some of their national research programs.

still important providers of information on technological options for agricultural development and the optimal use of resources.

In all cases, research, policy, regulatory, and linkage functions were closely intertwined. Functions other than experimentation proved to be very important for research systems in small countries, and in several cases, activities in science-technology policy and linkages appeared to be particularly important.

THE EVOLUTION AND DEVELOPMENT CYCLE OF NATIONAL RESEARCH SYSTEMS

The case studies traced the historical evolution of research systems, which is interesting for several reasons. First, the development cycle of national systems showed that even in those countries with long-established research systems, the research organizations have remained small, confirming the fact that our sample countries indeed face fixed constraints on the scale of their research systems (Box 4.2). Second, in tracing the historical origins of research in small countries, the role of export industries and producer associations was revealed to be very important. Third, while the scope of research in all the case-study countries has tended to expand over time, the institutional responses in each of the countries have been quite different. In some cases the expanded scope of research has led to moves to consolidate research in a single organization, while in other cases it has resulted in the establishment of new organizations to address new domains within the national research portfolio.

INSTITUTIONAL CULTURE AND LEADERSHIP

While the case studies focused on the structure and organization of the national research systems, several of them highlighted the importance of institutional culture and leadership in keeping small systems viable and effective. A strong institutional culture

---

Box 4.2. The Developmental Cycle of Research Systems

Two of the case-study countries have research systems that are larger than those of much larger countries. Mauritius and Honduras, for example, have over 100 researchers if one totals the scientific staff of all their institutes. This exceeds the number of researchers in larger countries such as Angola, Zaire, and Mozambique, which are at a low point in their developmental cycles, because of war or institutional collapse or both. However, the development needs of these countries clearly justify, and indeed require, more robust research systems, with several hundred researchers, to meet the needs of their agricultural sectors. The small size of their research systems is not related to the size of the country but to the temporary degradation of their economies and their national institutions.

Honduras and Mauritius are contemplating down-sizing their systems or devolving responsibilities for research to producers who can contract out research on an ad hoc basis. In the case of Honduras, a new law is now in effect to begin implementing the rescaling and restructuring of the system.* Mauritius, which has relied principally on national funding for its government, parastatal, and private-sector research, is now concerned about its ability to absorb and usefully employ external funding via World Bank loans. Growth in the system is not the goal; in fact, there is considerable feeling that public agricultural research services could be smaller and more focused. Parastatal research for the sugar industry is also not expected to grow, given the long-term decline in prices and in the relative importance of the industry. The concern in both Honduras and Mauritius is for research investments that build a flexible science and technology base, one that can meet a variety of objectives within agriculture and natural resources as well as across sectors.

*Establecimiento del la Dirección de Ciencia y Tecnologías Agropecuaria (DICTA), Acuerdo 2026-92, 7 October 1992. Executive Decree.

---

can build a sense of identity and commitment to an institution and its goals. This is crucial for national institutions that rely on external resources and projects to carry out their work. Institutional culture and leadership were essential ingredients in keeping organizations productive and in maintaining scientific credibility in times of instability and resource scarcity. Organizational values that emphasize service and focus on farmers are important in all research organizations concerned with agricultural development; however, the small size of the organizations involved here made it easier to instill such values. At the same time, when these values were not instilled, the tendency of scientists to pursue their individual objectives was even more pronounced.

There was no single structure, mechanism, or amount of money that secured loyalty and excellence. But two common, interrelated, themes surfaced: the importance of (1) leadership in defining a mission and (2) organizational flexibility in exploiting new opportunities and forging alliances to tap new sources of support. The more flexible organizations with clear accountability to policy and clients were also more conducive to the development of creative leadership.

HONDURAS: BUILDING ON DIVERSITY

Honduras produces a wide range of agricultural products, including traditional exports that are the country's principal source of foreign exchange. The need to improve rural incomes and diversify agricultural exports has led to efforts to find new crops and products that can be integrated into existing production systems and also make better use of underutilized resources and regions. In the face of these demands for increased agricultural productivity, the country's varied natural resource base, which includes temperate mountain forests, tropical rain forests, a variety of soils, and coastal resources, is rapidly being depleted. Honduras has a pressing need for research with a wide scope. However, as one of the poorer countries in Latin America, it has little chance of significant increases in funding or in the number of staff assigned to the existing research institutions (see Figure 4.1).

Despite this size constraint, Honduras has one of the more comprehensive agricultural research systems, with a greater capacity for technology generation than many small countries. How did it acquire this institutional capacity in research? Contreras (1992) argues that it is the result of institutional diversity—several distinct research organizations focusing on specific domains within the national research portfolio—and the strong participation of private-sector agroindustries. The key to improving the performance and efficiency of this research system is therefore to coordinate the existing research components and institutions around a national research policy that meets the needs of producers and development. To do this, public-sector research bodies need to play a more active role in the policy area.

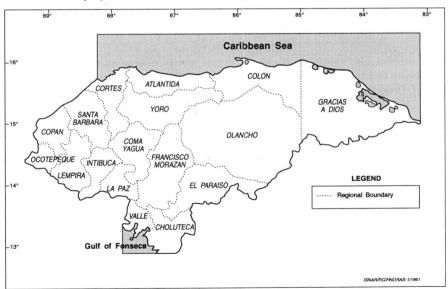

Figure 4.1. Honduras

POLICY ISSUES IN THE NATIONAL AGRICULTURAL
RESEARCH SYSTEM OF HONDURAS

Institutional comparative advantage: The Honduran NARS evolved from both development and technological needs as perceived by a variety of different actors, including private export companies, development agencies, government, and universities. The system is composed of centralized government research departments, parastatal commodity research institutes, semiautonomous government programs, private-industry research organizations, research programs in schools of agriculture and universities, and a private agricultural research foundation.

Table 4.2 shows the distribution of Honduras' research institutions across the national research portfolio. This is the basis for determining the individual comparative advantages of each institution for research in its respective domain (see Chapter 3). All of these agricultural research units address fairly specific commodities and clients, although overlapping does occur at times. One weakness of the system is the lack of an adequate mechanism for research policy orientation and support of the overall NARS.

Funding: The scale of agricultural research in Honduras is still modest, although it has expanded considerably during the last 10 years. The agricultural gross domestic product (AgGDP) for 1990 was estimated at about US$ 1,200 million in current values. Public-sector investment in agricultural research for 1990 was approximately US$ 5 million and represented 0.4% of the AgGDP. If we include public, private, and university agricultural research expenditures, the total averaged about US$ 8 million in 1989/1990 (0.6% AgGDP). Figure 4.2 illustrates the distribution of financial resources across the various NARS institutions and shows that the greater share of research expenditures is in private and parastatal research organizations. The two research departments of the Ministry of Natural Resources account for about 10% of the total research expenditure, while the banana multinationals and Fundación Hondureña de Investigación Agrícola (FHIA) account for over half. The two parastatals, Centro Nacional de Investigación Forestal Aplicada (CENIFA) and Instituto Hondureño del Café (IHCAFE), account for nearly 25% of the total expenditures. The key to getting the most out of this diversified portfolio of research investments is improved coordination at the policy level.

Research policy and planning: In practice, each institution defines its research agenda. There is no operational or effective mechanism in Honduras to review the programs of the various institutions either in relation to the work that others are doing or in relation to the overall national investment in research and its priorities. Other developing countries have created national bodies or councils specifically to formulate research policy and to coordinate a diverse set of institutions around national goals. Honduras is now beginning the implementation of one such mechanism, the Dirección de Investigación y Tecnología Agropecuaria (DICTA), which will promote decentralization and private-sector participation.[3]

Table 4.2. Research Components and Research Scope in Honduras, 1990

	National Research Portfolio						
NARS Institutions	Global Staples	Trad. Exp. Crops	Minor Food Crops	High-Input, Nontrad., Exp. Crops	Livestock	Socioecon. & Rural Engineer.	NRM
MNR							
a. DIA	Beans Maize Potatoes Rice Sorghum						
b. DIP					Beef cattle Dairy cattle Feeds & nutrition Pastures		Range management
EAP	Sorghum			Melons		Farming systems res.	Integrated pest man.
IHCAFE		Coffee				Marketing	
a. Standard Fruit		Bananas		Pineapples Grapefruit			
b. Chiquita Brands		Bananas Oil palm					
FHIA		Bananas Cocoa	Onions Peppers Plantain Tomatoes	Black pepper High-value vegetables Mangoes		Marketing Socio- economics	
CENIFA						Social forestry studies Wood uses	Forestry

MNR:	Ministry of Natural Resources.	IHCAFE:	Instituto Hondureño del Café.
DIA:	Departamento de Investigación Agrícola.	FHIA:	Fundación Hondureña de Investigación Agrícola.
DIP:	Departamento de Investigación Pecuaria.		
EAP:	Escuela Agrícola Panamericana.	CENIFA:	Centro Nacional de Investigación Forestal Aplicada.

Regional and international cooperation: The prospects for stronger regional and international cooperation in research is another important research policy and management issue. In Central America, there are export crops, staple food crops, and natural resource management problems that are common to several countries in the region. There are important regional research activities for global staples such as maize, beans, and potatoes that are supported by the research programs of the international centers. Instituto Interamericano de Cooperación para la Agricultura (IICA) is also a key supporter of collaborative regional programs.

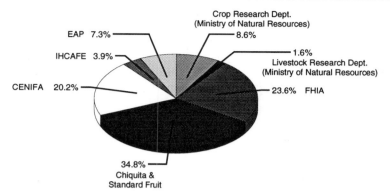

Figure 4.2. Distribution of research expenditure by research institution in Honduras, 1990

In addition, there are opportunities for national research programs to collaborate on traditional exports, such as coffee,[4] and the nontraditional exports that have become established in the region, such as melons (Contreras 1992). The types of linkages needed to collaborate and tap external sources varies according to the technology domain. Because these are strategically important commodities in a competitive world market, policy-level coordination is important if countries are to collaborate in research.

Human resource planning and accountability in a small NARS: The number and quality of national research personnel in the country has increased significantly over the last two decades (Table 4.3). However, high staff turnover in government research services has resulted in the loss of useful information and research results. Some programs are left unfinished or repeated as scientists leave or are reassigned. Another problem for researchers in small countries is that opportunities for peer review and publication are limited. Special mechanisms, including opportunities for in-service training, contracting services, and study travel, may encourage stability in staffing and help overcome problems of scientific isolation. Some of these incentives, along with mechanisms for monitoring and evaluation, can increase the accountability of researchers to their clients. With the exception of research programs in the banana industry, there has been little accountability to users.

Staffing profiles may be different in small countries where large teams of scientists are unlikely to be working in depth on a single problem. In the past there was a tendency to concentrate training on a few narrow subject areas with few opportunities for interdisciplinary training. The focus was often on plant breeding; other fields such as agronomy, postharvest technologies, and plant and animal physiology were neglected. Generally speaking, research systems in small countries have had difficulty in making effective use of highly trained specialists with a narrow focus. Highly trained scientists are needed, but they should have a broad focus to be able to cover several commodities and the flexibility to change their focus over time as needed.

Table 4.3. Research Institutions in Honduras: Human and Financial Resources, 1990

Institutions	Researchers				Expenditures
	Total	PhD	MSc	BSc	
Departamento de Investigación Agrícola, Ministry of Natural Resources	69^a	2	12	55	US$ 712,910 (3,037,000 Lempira)
Departamento de Investigación Pecuaria, Ministry of Natural Resources					US$ 121,125 (516,000 Lempira)
Fundación Hondureña de Investigación Agrícola (FHIA)	29	9	9	11	US$ 1,800,940 (7,672,000 Lempira)
Research Department, Standard Fruit Company	4	–	–	–	US$ 2,800,000
Research & Technical Department, Chiquita Brands	5	–	–	–	
Centro Nacional de Investigación Forestal Aplicada (CENIFA)	35	2	15	18	US$ 1,629,550^b
Research Department, Instituto Hondureño del Cafe (IHCAFE)	21	0	5	16	US$ 300,000 (1,278,00 Lempira)
Other Institutions					
Escuela Agrícola Panamericana (EAP) (Integrated Pest Management Program)	30^c	5	9	16	US$ 590,000^d
Total	*193*	*18*	*50*	*116*	*US$ 8,048,550*

Note: The rate of exchange is based on 4.260 Lempiras/US$ in 1990 (World Bank 1995).
a.   Apparently includes nonresearch staff of the ministry.
b.   Expenditure is for 1989.
c.   Includes all professional staff. Full-time equivalents in research estimated at seven person-years.
d.   Funding of the Integrated Pest Management Program only.

GAPS AND EMERGING DEMANDS ON HONDURAS'S NATIONAL RESEARCH PORTFOLIO

In general, the technological needs of the transnational corporations have been addressed effectively by their research programs in Honduras. This includes bananas and other export commodities of interest to those companies. Their proven ability to adapt to marketing and production conditions points to continued development and adaptation of agricultural technology to meet their needs.

Unfortunately, this is not the case for staple food commodities. Government and, more recently, private research programs have not been able to develop technological solutions to the production and postharvest constraints of the resource-poor farmers who comprise most of the farming community but who contribute only a small part of AgGDP. Research for subsistence farmers poses a far more difficult ecological, economic, and social challenge. Recent measures to privatize research even further should be accompanied by special policy incentives so that research is aimed at food security and improving the welfare of the rural poor; otherwise, these objectives could easily be ignored under private research schemes.

There are also notable gaps in the current national scope of research. Relatively little attention has been given to animal production, both commercial and noncommercial, although livestock accounts for a sizable share of agricultural output. In addition, there is very little research on fish and shrimp aquaculture, which are emerging as important contributors to food security and as major high-value export commodities for the country. Late in 1991, the Dirección General de Pesca y Acuacultura (DIPESCA) was created within the Ministry of Natural Resources to provide technical support, including research, to establish resource management and conservation policies for the Honduran shrimp, lobster, and fish industries (Zacarías 1992).

Finally, there is a need for greater input from research on natural resource management policy for the conservation and use of the forest resources that cover over two-thirds of the country. Agricultural activities often lead to greater deforestation and competition within the natural resource sector. Honduras must find a balance among agriculture, forestry, and conservation if it is to sustain agricultural and forestry productivity into the future. Extending the scope of research to cover the breadth of topics in natural resource management is urgent but difficult for existing agricultural research organizations to handle.

Research on natural resource management in Honduras is conducted largely within university research programs, as in the case of integrated pest management at the Escuela Agrícola Panamericana or the forestry research in CENIFA. This strategy has enabled Honduras to extend natural resource management (NRM) research to include watersheds, wildlife, and environmental policy. As this research is long term and knowledge based, universities may provide the most appropriate environment for combining the generation of income from commercial forestry with the long-term sustainability of Honduras's varied natural environment.

Recent government policies are aimed at privatizing research programs. The rationale is to improve the efficiency, relevance, and accountability of research. The final outcome of this move depends heavily upon the ability of the private sector to sustain the research programs on strategically important commodities, while fulfilling the national policy objectives concerned with equity and food security. Without consistent public support, the country risks losing the human and physical infrastructure that has been built to address staple food production issues over the last 30 years. Without research policy guidance and coordination, the traditional export commodities may receive an even larger share of resources, while other areas may be neglected and equity and environmental concerns may be deferred. The diverse research organizations in Honduras have put together the essential blocks for an effective national system. Now they must be placed into a workable, sustainable structure and policy framework.

JAMAICA: ADJUSTING TO CHANGING PRIORITIES

Jamaica's national research system has three distinctive, and instructive, features (see also Reid 1992, upon which this section is based). The first is the decline in the size of the parastatal commodity research institutions. The second is the rise of new, more flexible mechanisms to provide incentives for producers and private companies to invest in research. And the third is Jamaica's participation in the regional research programs executed by the Caribbean Agricultural Research and Development Institute (CARDI). The importance of research policy to manage institutional change, to promote greater flexibility in order to respond to rapidly changing markets, and to coordinate a variety of national and regional actors in research, emerges as a clear finding in this case study (Reid 1993).

THE SCALE AND SCOPE OF THE NATIONAL AGRICULTURAL RESEARCH SYSTEM

Jamaica has many institutions involved in agricultural research and development, but the absence of policy guidance and coordination has not allowed them to function as a coherent system. At present, there are 19 national, regional, and intergovernmental organizations (public, parastatal, and private) that are wholly or partly involved in agricultural research. The Research and Development Division within the Ministry of Agriculture has the broadest scope: it addresses 27 different commodities. The

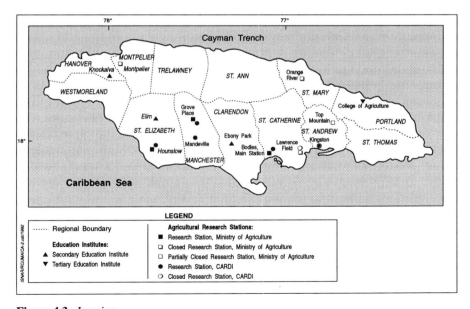

Figure 4.3. Jamaica

USAID-supported Jamaica Agricultural Research Project (JARP) covers 18 different areas, and CARDI's Jamaica program works on an additional 16 topics and commodities. Except for sugar, the focus of the commodity organizations is on tree crops, with some interest in commodity-based intercropping systems (e.g., pasture or cocoa with coconuts, coffee and bananas with cocoa). In the case of sugar, new commodities suitable for intercropping are being assessed as potential contributors to a diversification program. Table 4.4 illustrates Jamaica's national research portfolio.

Some institutions not primarily involved in research or agriculture are making significant contributions to the system. The Scientific Research Council (SRC), for example, provides a vital link between commodity and food research and agro-processing and postharvest technologies, including the use of indigenous raw materials in industry. Private-sector companies do some research to identify options for technology transfer in order to solve problems within existing commercial production systems.

Based within the recently created nonprofit Jamaica Agricultural Development Foundation (JADF), JARP's goal has been to increase the long-term involvement of the private sector in research by directly funding areas of interest to both private companies and the public. It has also been able to fund noncommercial research, including a program on the conservation of biodiversity. Beneficiaries of JARP funding include the University of the West Indies (UWI), CARDI, and special private-sector concerns as well as the public sector and parastatal bodies (Wilson 1992).

Jamaica has called on CARDI's regional research capacity to fill crucial gaps in the research scope and to intensify national research in areas that are important to the region of the Caribbean Community and Common Market (CARICOM) as a whole. Accounting for approximately 20% to 25% of the human and financial resources available to Jamaica's research, CARDI research is oriented towards producing "public goods" that can be transferred and adopted in other CARICOM countries. Linkages with regional programs run by CARDI can potentially benefit Jamaica in 28 relevant areas of research.

UWI has 19 professionals who are active in research, eight hold PhDs and two have Masters degrees. There is no research budget per se that can be directly attributed to research because teaching and research duties are not segregated and most research is done under contract to other agencies in the NARS, which manage research budgets. Nonetheless, UWI staff have contributed to agricultural research in 15 areas of relevance to the national research portfolio (see Table 4.5).

RESEARCH POLICY ISSUES AND SETTING PRIORITIES

Changes in Jamaica's agricultural economy and its relation to the mining and tourism sectors pose immediate challenges for research policy. Jamaican agriculture is still searching for commodities where it may have a comparative advantage. Traditional exports such as sugar, coffee, and bananas are facing declining prices and growing

Table 4.4. Jamaica: Research Components and Research Scope, 1991

				National Research Portfolio			
NARS Inst.	Global Staples	Trad. Exp. Crops	Minor Food Crops	High-Inp., Nontrad. Exp. Crops	Livestock	Soc. & Rural Engineer.	NRM
MOA (R&D Division)	Maize Sorghum	Coffee	Pumpkins Yams Swt potatoes Pigeon peas Beans (red pea) Onions Swt peppers Hot peppers Plantain Cabbage	Cucumbers Citrus Ornamentals Passion fruit Papaya Broccoli	Pasture Nutrition Animal breeding Animal health Cattle Goats Sheep		Germplasm Plant protection
Banana Board		Bananas					Plant protection
CaIB		Cocoa					Plant protection Soil cons.
CoIB		Coconuts					Plant protection Soil cons.
CIB		Coffee				Marketing research	Plant protection Soil cons.
CGA				Citrus		Irrigation	Plant protection Soil cons.
SIRI		Sugar				Machinery & tools Prod. sys.	Gen. res'rce conserv. Plant protection Soil cons. Water res'rce mngt.
SRC		Bananas			Goats Rabbits	Agroproc. Market research	Fisheries Culturing of algae
SPID						Postharvest & storage	

Note: CARDI (Caribbean Agricultural Research Development Institute) and UWI (University of the West Indies) are not part of Jamaica's national research portfolio.
CaIB = Cocoa Industry Board; CoIB = Coconut Industry Board; CIB = Coffee Industry Board; CGA = Citrus Growers Association; SIRI = Sugar Industry Research Institute; SRC = Scientific Research Council; SPID = Storage and Prevention Infestation Division.

Table 4.5. Research Institutions in Jamaica: Human and Financial Resources, 1991

Institutions	Researchers				Expenditures
	Total	PhD	MSc	BSc	
Department of Agricultural Research, Ministry of Agriculture & Commerce	27	2	7	18	US$ 1,066,115 (J$ 12,9000,000)
Sugar Research Institute, Jamaica Sugar Industry Board	18	–	4	14	US$ 749,008 (J$ 9,063,000)
Research & Extension Department, Cocoa Industry Board	3	1	1	1	US$ 53,719 (J$ 650,000)
Research Department, Jamaica Banana Board	3	1	0	2	US$ 82,644 (J$ 1,000,000)
Research Unit, Jamaica Coconut Industry Board	4	0	2	2	US$ 183,471 (J$ 2,220,000)
Research Unit, Coffee Industry Development Co.	4	0	2	2	US$ 19,018[a] (J$ 120,000)
Other Contributors					
Jamaica Agricultural Development Foundation	NA	–	–	–	US$ 1,713,091[b] (J$ 12,300,000)
Scientific Research Council	23	4	8	11	NA
CARDI	18	4	5	9	US$ 760,661 (J$ 9,204,000)
Total	100	12	29	59	US$ 4,627,727

Note: The rate of exchange is based on J$ 12.1/US$ for 1991 (World Bank 1995).
a.   Expenditure is for 1988.
b.   Expenditure is for 1990 (exchange = J$ 7.18/US$ for 1990 [World Bank 1995]).

competition. The promise of high-value nontraditional exports that would take advantage of "windows of opportunity" in northern markets has been oversold despite the support of programs such as the Caribbean Basin Initiative (see Chapter 6).

Growing concerns with environmental impact and sustainability also need to be included in national research planning. Among the critical areas needing research attention are environmental protection (including restoration), soil and water conservation, fertility management, integrated pest management, postharvest techniques, livestock production systems, and by-product utilization. Jamaican agriculture has increasingly had to plan its development in relation to other natural resource sectors and other industries such as tourism. Socioeconomic factors, including marketing and employment, are now central to the objectives of agricultural research and development.

In Jamaica, as in the other countries studied, the ultimate responsibility and authority for coordination, monitoring, and evaluation is properly with the Ministry of Agriculture. The relevant sections within the ministry may need to strengthen their capacity to provide policy guidance, which may be a more important function than implementing research programs. A clear research policy would enable the various actors to set their priorities in accordance with their institution's particular comparative advantage and their contribution to national development goals.

Jamaica's fragmented institutional and subsector planning has made it difficult to focus on areas with future prospects and to solve the long-term problems that have resulted from inefficient use of natural resources. Furthermore, it is difficult to reallocate resources from areas with declining prospects to those with future importance.

Jamaica is searching for priority-setting methods that fit the size and scope of its agriculture and its natural resource sector. One finding of this study has been that general guidelines and indicators on levels of investment are difficult to apply in a small country with an open export economy. These formulae do not account for the fact that there are different types of research responses that can address similar needs. For example, a very important crop can be well covered simply by obtaining research results from countries with larger research programs. A new commodity with great potential may not appear important, based on current production values, but it could well merit greater research investments because of its promise and the fact that relevant research results are not available elsewhere. For example, it is misleading to say that Jamaica should allocate 1% of the value of a commodity to research. That would apportion more to sugar than it possibly needs while ignoring more promising commodities where the R&D costs are lower. The "1% per commodity" congruence model would have little impact in the new area of nontraditional exports, for example.

Scientists and policymakers still prefer general and simple guidelines, such as (a) focus on the most important researchable problems and commodities, (b) aim for excellent quality in output, and (c) establish and maintain strong information linkages. How to translate these goals and guidelines into decisions on resource allocation, however, is not clear, but a more formal and broad-based priority-setting process is essential.

In practice, priority-setting methods need to assign more objective values and weights to economic, political, social, and scientific criteria, as well as including environmental criteria. Priority setting in Jamaica also has to take regional research into account, given the crucial role of institutions such as CARDI and UWI. The general, simple guidelines of yesterday need more robust indicators and criteria in order to cope with new, more complex demands and more diverse objectives. Jamaica has built a research system upon a diverse set of institutions, which is good; however, the system still lacks the planning and priority-setting mechanisms it needs to give it the flexibility to make the most of this diversity and to cope with change.

FIJI: REBUILDING TO ASSUME REGIONAL RESPONSIBILITY

In the South Pacific, Fiji, with its population of 1.3 million, is one of the few countries that has a national research system with institutes organized around traditional experimentation and commodity-improvement functions. In a region of very small island states, most of which are not able to sustain technology-generation programs, Fiji has an important role to play. Many of the crops that are crucial to the diet of the

South Pacific peoples, such as taro, sago, sweet potatoes, and breadfruit, are "or-phans" in world trade and in global research and development. Fiji's national research can be an important source of technology and information on these and other staple food crops in the region, and assuming these regional responsibilities presents an additional challenge for this small national system (Sivan 1992a).

In addition to Fiji's important agroindustries, such as sugar, coconuts (to a lesser extent), and, increasingly, silviculture and forest products, the population density on the island has created a demand for staple food crops and local horticultural products. Agroindustry and population pressures have also put pressure on the system for research on the management of natural resources to cope with pollution and resource depletion.

THE SCALE OF THE NATIONAL AGRICULTURAL RESEARCH SYSTEM

Fiji's research system is publicly owned and is largely managed directly by govern-ment ministries (for more information on Fiji's system, see Sivan 1992a,b). Agricul-tural research is carried out in four divisions or sections under two government ministries and in two parastatal organizations. Within the Ministry for Primary Industries and Cooperatives, there is a research division responsible for crop and soil research. The Animal Health and Production Division and the Fisheries Division of the same ministry each have their own research units.

Forestry research is carried out in the Ministry of Forestry. To promote large-scale development of pine forests, the Fiji Pine Commission, an autonomous government

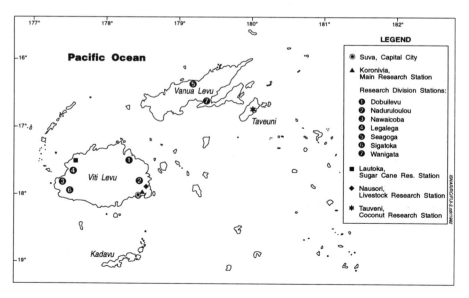

Figure 4.4. Fiji

organization, was formed and it contains a small research section for pine. Fiji's sugar industry was developed by a private company based in Australia, the Colonial Sugar Refining Company (CSR), which established a sugarcane research station at Lautoka in 1912. After independence in 1970, the Fiji government bought all the sugar operations, including research, from CSR and formed a parastatal, the Fiji Sugar Corporation, to manage its activities.

The Fiji case shows how dramatic changes in the political environment affect the institutional stability of research. The existence of two major cultures, native Fijians and Fijians of East Indian origin, has led to political struggles over the allocation of resources within the government. Following the coup of 1987, the new government pressured government agencies to give preference to native Fijians who had formerly been underrepresented in government services and education. While the need to increase educational and professional opportunities was clear, these rapid changes led to the loss of many experienced researchers.

Table 4.6 shows the total budget and the number of scientists in the different research institutions over the last decade. These grew steadily in the early 1980s and peaked in 1986. After the coup in 1987, both the budget and number of scientists decreased. In early 1987 there were 27 research scientists in the research division, but this number had declined to 17 in 1990.

Table 4.6. Research Institutions in Fiji: Human and Financial Resources

Institutions	Year	Researchers				Expenditures	
		Total	PhD	MSc	BSc	Year	Amount
Research Division, Ministry of Primary Industries	1990	12	3	5	9	1991	US$ 2,027,000 (F$ 3,000,000)
Research Section, Animal Health & Production, Ministry of Primary Industry	1989	5	0	1	4	1989	US$ 510,666 (F$ 766,000)
Sugarcane Research Institute, Fiji Sugar Corporation	1990	6	0	1	5	1987	US$ 833,000 (F$ 1,000,000)
Silviculture Section, Ministry of Forestry	1987	1	–	–	–	1987	US$ 240,000 (F$ 288,000)
Pine Section, Ministry of Forestry	1987	1	–	–	–	1987	US$ 63,000 (F$ 76,000)
Timber Utilisation Section, Ministry of Forestry	1987	4	–	–	–	1987	US$ 173,000 (F$ 208,000)
Other Contributors							
Fisheries Division, Ministry of Primary Industries	1989	1	0	0	1	1987	US$ 361,000 (F$ 433,000)
University of the South Pacific		NA	–	–	–	–	NA
Total		30	3	7	19		US$ 4,244,000

Note: Data are for the most recent year available at the time of the study.
Exchange rates are F$ 1.48/US$ for 1991, F$ 1.5/US$ for 1989, and F$ 1.2/US$ for 1987 (World Bank 1995).

Now the system is slowly rebuilding its staff and programs, which presents the opportunity to build more flexible and better integrated structures that can respond to the needs of Fiji's subsistence sector as well as the changing fortunes of the agroindustrial and natural resource sectors.

The Research Division in the Ministry of Primary Industries was better developed than the other divisions and has had a stable base of funds from national sources. In contrast, the livestock and fisheries research units are more dependent upon external project funding; the weak and rather unstable support that their research receives does not reflect the importance and actual potential of these two sectors. Integrating livestock and fisheries research into the research division might give them more visibility and better funding.

SCOPE OF RESEARCH

Apart from the Sugar Cane Research Centre (SRC), which conducts some basic and applied research, other institutions in Fiji are almost entirely engaged in research at the adaptive and testing levels. The SRC is organized on a disciplinary basis with programs in breeding, agronomy, physiology, plant protection, and plant nutrition. The ministry's research division is currently working on rice, maize, cocoa, coconuts, root and tuber crops, vegetables, fruits, and grain legumes, as well as maintaining disciplinary programs in plant nutrition, plant protection, and agricultural engineering. With its broad research mandate, it also maintains a watching brief on research problems and technology in some of the export crops, both traditional and nontraditional, not covered by other institutions. The Livestock Research Section works on dairy cattle, beef cattle, goat, sheep and swine production; pasture improvement; and seeds. The Forestry Research Section conducts production-oriented research on silviculture, pine, and hardwood production, and timber utilization. The Fisheries Research Section is more focused on natural resource management and has programs in aquaculture, giant clam production, and marine resource assessment and development. Table 4.7 presents the national research portfolio of the country.

Increasing the scope for research: In the Research Division some economies of scope have been made by researchers working across a number of commodities. With better coordination, it would also be possible for researchers to work with the programs of other institutions, such as pasture and animal feed research in the Livestock Research Section. Organizational changes that facilitate the movement of staff, knowledge, and resources across existing components is one way of extending the scope without major increases in the scale of operations.

The scope for research can also be increased by improving the linkages with external institutions and obtaining a range of technologies for testing under the various ecological conditions in the country. Currently, Fiji's sources of technology and information are fairly narrow and are determined largely by historic ties to other countries in the Pacific, such as Australia, and the commodity programs of the

Table 4.7. Fiji: Research Components and Research Scope, 1990

				National Research Portfolio		
NARS Institutions	Global Staples	Traditional Export Crops	Minor Food Crops	High-Input Nontrad. Export Crops	Livestock	NRM
Research Division, Dept of Agriculture	Beans Cowpeas Goundnuts Maize Potatoes Rice	Bananas Cocoa Coconuts	Cabbage Carrots Garlic Onions Pigeon peas Taro Tomatoes Vegetables	Cardamom Citrus Ginger High-value vegetables Mangoes Papaya Passion-fruit Pineapples	Goats/ sheep Pasture/ fodder Swine	
Research Section, Animal Health & Production Division					Cattle Goats Sheep Swine Pasture/ fodder	
Sugar Cane Research Centre		Sugar				
Silviculture Section, Ministry of Forestry						Forestry Forest protection
Pine Section, Ministry of						Forestry Agrofor. Forest protection
Timber Utilisation Section, Ministry of Forestry						Processing/ postharvest Storage & transport
Fisheries Division, Ministry of Primary Industries						Fisheries

Note: Socioeconomics and rural engineering are not included in Fiji's research portfolio.

international centers. Donor projects also provide important linkages to external sources, but these are often not institutionalized or accessible to the larger R&D community in the country.

For some of Fiji's important crops, such as taro, improved technology is not readily available from abroad: applied taro research has to be done locally. A breeding project was started in Fiji which successfully developed some high-yield cultivars; however, this program was dependent on one scientist. To keep the effort going, the Institute for Research, Extension, and Training in Agriculture (IRETA) was able to assist in evaluating the improved clones coming out of the program and brought the program to a successful conclusion.

There are two major conclusions that can be drawn from this. First, if relevant technologies are not available from outside for a major commodity within a small nation, there is little choice for the country but to develop them themselves. Second, it is difficult for a small country to sustain long-term research for generating new technologies. Where a crop is of importance to a number of small countries, some of the research, such as breeding, could be done on a regional basis. Based at the University of the South Pacific, IRETA has now developed an extensive breeding project for taro in the region. National research supported by a regional program may be one way to provide stability for research on crops and agricultural problems specific to small Pacific island nations. Given the strategic role that Fiji's NARS could play in regional food crop research, more could be done at the policy level to win support for national research as a way of building regional research capacity on Pacific food crops and livestock.

NONRESEARCH FUNCTIONS

One finding from the Fiji case study that is relevant to most small countries is the extent to which research organizations are expected to perform a host of other development and regulatory functions. For example, in Fiji, the Research Division is required to provide the following nonresearch services:

- seed production: producing rice, vegetable, and maize seed and planting materials for root crops and tropical fruits for sale to farmers and commercial producers at a subsidized cost;
- seed testing: providing a seed testing service to the public;
- soil and plant analysis: providing soil and plant analysis services for the ministry and the general public;
- regulatory services: registering pesticides and regulating the use of pesticides in the country;
- advisory services: providing advisory services for quarantine and pest and disease control;
- food and forensic analysis: providing analytical services for food and forensic materials to the Ministry of Health and the police.

During 1987, six scientists of a total of 27 full-time researchers in the Research Division were providing these services. Other research institutions in the country, such as the livestock and pasture research section, raise and provide animal breeding stock and pasture seeds for farmers.

Some activities with no relation to agriculture, such as the services for food and forensic analysis, should be separated and put under a national laboratory. However, other services linked to agricultural production would be more difficult to separate from research in a small country like Fiji where there are no other organizations to carry them out. The research stations have the resources and the research staff as well as the technical know-how to supervise seed production, conduct seed testing, do soil and plant analysis, and provide services for pesticide registration and plant protection. It is very difficult and expensive for small countries like Fiji to set up separate organizations to provide these services. NARS institutions need to adopt strategies to cope with multiple research and nonresearch functions such as these.

LINKAGES

The Research Division and the Sugar Cane Research Centre have good links with external sources of information and technology. However, many of these were developed by personal contacts and initiatives and were then lost when the staff left. While individual researchers are often the main transmitters and storehouses of information in a small research system such as Fiji's, additional mechanisms are needed to make sure that relevant information is institutionalized and retained.

There is very little local support for developing and maintaining links with external institutions that are identified as relevant sources of knowledge and technology; most of the international meetings, workshops, conferences, etc., that staff attend are funded by outside organizations. These external sources of technology and information need to be considered in determining the scope and level of national research programs. Hence, the importance of managing these information links.

Linkages with users: The Research Division has established mechanisms to receive inputs from users, extension agents, and farmers in the program review process. Other research units have less formal links since they are in fact embedded in development divisions. Despite these measures and the fact that research and development units exist within the same organization, the linkages between research, technology transfer, and users are weak and various recommendations have been made to strengthen them. One option is to create a technology-transfer unit within research that can translate research information into simple language for extension staff and farmers and to organize training and media demonstrations. Such a technology-transfer unit was developed as part of a rice research project funded by Japan and was very effective in training extension staff and farmers. There was interest to extend the function of the unit to cover other commodities; however, this was not possible due to a lack of additional donor support.

One dilemma is that developing formal research extension units also creates additional burdens on small organizations with few staff and limited budgets, and relying on donor projects is an unstable option. Perhaps integrating the farmer into the research process through new methodologies is a more useful approach for small research systems. This could exploit the advantages of smallness in reducing the distance between researchers and farmers. Farmers already go directly to some researchers for advice, and whether this is seen as a drain on a researcher's time or an opportunity for the researcher to get feedback on his or her research is largely a matter of method and perspective.

RESEARCH POLICY, PLANNING, AND MANAGEMENT

Research on agriculture and natural resources is carried out in six different institutions: one division and three sections in two ministries and two parastatal organizations. There are many obvious areas of overlap, such as that between crop and soil scientists from the Research Division and pasture scientists from the Livestock Section in plant nutrition and plant protection.[5] In agroforestry, there is a need for close cooperation among all scientists in crop, soil, livestock, and forestry research because Fiji's farmers integrate crops, livestock, and trees in their production systems. There is also substantial production of traditional food crops and livestock in sugarcane areas, and competition between these commodities for land is intense. These factors underline the need for an integrated approach to agricultural research across the existing divisions.

There is no formal mechanism to coordinate research among these institutions. Ad hoc committees, comprising research and extension personnel from government ministries and the sugar industry, have been formed in the past to coordinate research and production, but they have been largely ineffective. Coordination needs to be at a higher policy-making level in order to put forward strong, coordinated support for research.

The Ministry of Primary Industries has developed a long-term plan for the Research Division. It sets out programs and priorities for research, and reviews and monitors the division's activities. However, there were no such plans or review procedures for the other research institutions at the time of this study. There is a need for a long-term research plan for all institutions as well as a formal mechanism to review, monitor, and adjust the programs annually. A consolidated mechanism that covers all the research organizations in the natural resource sector, including fisheries and forestry, is needed to develop national research policy. This is something that must be considered by policymakers and research managers if the effectiveness and efficiency of the research system is to be improved.

A mechanism that has proved useful in other countries is a national council for research consisting of senior government officials, research directors, and private-sector personnel with the power to monitor, direct, and coordinate research. This option requires government to devolve certain research policy functions to this body. An alternative would be a national advisory committee on agricultural research with similar membership but which would only act as an advisory body on research to the government.

HUMAN CAPITAL: THE KEY TO FUTURE DEVELOPMENT

Given the loss of staff over the last six years, Fiji has both the task and the opportunity to create a new system. Research leaders and policymakers need to decide on the scale of the system they need and can sustain. Given that scale, they can then decide on the scope of research, upon which domains to concentrate, and what approaches to follow. From that, Fiji's research leaders can determine the human resources that will be needed. Fiji is one of the two small countries in the South Pacific (the other is Papua New Guinea) that can sustain a research system of between 50 to 80 researchers spread over two or three research organizations in the public and parastatal sectors.

Forestry is a fast-growing sector, and the implication is that stronger links between crop, livestock, and forestry research will be needed. An important outcome might be a land-use policy and information system of use to all branches of natural resource research in Fiji.

On the basis of our study, we would argue that Fiji is capable of doing some applied research in root and tuber crops and in livestock. To justify the concentration of effort needed to do this type of research means taking on regional responsibility as a source of technology for root and tuber crops and South Pacific farming systems. In assuming the regional responsibility for key areas of technology generation and in coping with more complex natural resource approaches, Fiji will not only need to train the people to carry out this work, it will also have to sustain a strong and sophisticated national science capacity in agriculture and natural resources. Although this task must also be supported by regional partners, a stable national system in Fiji is crucial for research in the South Pacific region as a whole.

MAURITIUS: DIVERSIFYING RESEARCH INTO THE NEXT CENTURY

In 1993, Mauritius celebrated its centennial of organized agricultural research. During this time the country has relied upon relatively small institutions funded almost entirely from national sources and staffed overwhelmingly by national scientists (Manrakhan 1992). The experience of Mauritius demonstrates that a small country can build and maintain an effective agricultural research system based on local staff and resources. It provides a useful illustration of the conditions needed to institutionalize excellence in research.

The key to the institutional development of research in Mauritius has been the dominance of a single commodity—sugar—upon which the research system was built. The needs of the highly competitive sugar industry created a demand for scientific agriculture and research to develop improved technologies and practices. The profits from the sugar industry in turn financed the creation of research and training institutions. These sugar-based institutions have since diverged and diversified, resulting in the current research system. At its core are the Mauritius Sugar Industry Research Institute (MSIRI), the Scientific Services of the Ministry of Agriculture, Fisheries and Natural Resources (MAFNR), and the University of Mauritius (see Figure 4.5). Although the size of the national research system is not expected to grow dramatically in terms of numbers of scientists, the scope of research is changing and becoming more complex.

Manrakhan (1992) examined how a research portfolio focused mainly on sugarcane has been successfully broadened to encompass food crops, livestock, and nontraditional exports while paying increasing attention to conservation of the environment. The need for policy and coordination of this increasingly complex research system has been met by the creation of the Food and Agriculture Research Council (FARC) that promotes partnerships between the various research actors in the government, the university, and the private sector (Antoine and Persley 1992).

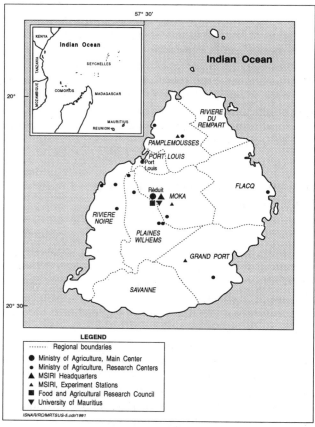

Figure 4.5. Map of Mauritius and the location of its research stations

OVERVIEW OF THE SCALE AND SCOPE OF THE SYSTEM

The greatest expertise and the most developed facilities for applied agricultural research aimed at generating new technologies are concentrated at MSIRI. The institute's focus is on sugar and commodities such as maize, beans, groundnuts, potatoes, and tomatoes that can be produced in association with sugarcane. This commodity focus is what has enabled MSIRI to develop the critical mass of scientists and resources to build multidisciplinary teams and assign them to long-term programs designed to produce new technologies. In generating technology, MSIRI attempts to apply new techniques, such as biotechnology, to produce the next generation of technologies that are essential for the future competitiveness of the Mauritian sugar industry. Since the sugar industry is the major user of agricultural land and water, concerns about natural resource management have become increasingly important within its research programs.

Although the Sugar Research Institute is the best-equipped crop research establishment for generating technology, the Ministry of Agriculture, Fisheries and Natural Resources has the greatest concentration of agricultural research capacity with the broadest scope. It is involved in adaptive research on the improvement of crop and livestock production, in the prevention and control of plant and animal diseases and pests, and in other aspects of agricultural development, including extension. The research is dispersed across the departments and agencies within the ministry. In terms of R&D, the ministry is well ahead of the other institutions discussed here, in terms of numbers of staff and funding inputs. But in terms of actual full-time-equivalent research, the gap between MSIRI and the ministry narrows considerably.

The close integration of research with extension and development services within the ministry poses problems for identifying research inputs and outputs. However, this integration is useful for bringing a comparatively small scientific effort to bear on a very wide range of topics. Research done in conjunction with producers, extension, and development may be a more effective way to conduct the screening, testing, and adaptive research that is the main focus of the ministry's scientific services. The following discussion of the scale and scope of research in the ministry considers the feasibility of such broad coverage. In light of the objectives of research, the level of activities needed for technology screening, testing, and adaptive work, and the integration of research with development and extension, the current scope may not be as unrealistic as it appears.

There are 13 crop and livestock research stations and centers within the Ministry of Agriculture, Fisheries and Natural Resources. There are also two fisheries centers: the Albion Fisheries Research Centre and the La Ferme Fish Farm. Three centers make up the Forestry Station. Natural resource conservation programs within the ministry also maintain deer paddocks, bird sanctuaries, and nature reserves. Most staff time and R&D expenditures at the ministry are devoted to development. Of the 93 full-time-equivalent scientists on the ministry's R&D staff, there are about 65 with

Table 4.8. Research Institutions in Mauritius: Human and Financial Resources, 1991

Institutions	Researchers				Expenditures
	Total	PhD	MSc	BSc	
Mauritius Sugar Industry Research Institute (MSIRI)	*55*	7	26	22	US$ 2,857,000 (44,713,000 Rs)
Agriculture, Livestock, Forestry & Scientific Services (MAFNR)	*73*	11	–	25	US$ 2,875,400 (45,000,000 Rs)
Fisheries Division and Albion Fisheries Research Centre (MAFNR)	*20*	–	–	–	–
Food and Agriculture Research Council (FARC)	*3*	–	–	–	US$ 639,000 (10,000,000 Rs)
Other Contributors					
Schools of Agriculture, Science, Engineering, and Agriculture (University of Mauritius)	*35*	–	–	–	No specific research budget
Total	*186*	18	63	47	*US$ 6,371,400*

Note: Rate of exchange based on 15.65 Rs/US$ (World Bank 1995).
Private-sector firms with research units include Food and Allied Industries Ltd., Camaron Hatchery Ltd., Camaron Production Co., and the Horticultural and Ornamental Producers' Association.

major research responsibilities. Even this number is not exclusively concerned with research. The ministry spends close to US$ 3 million per year on research and development, most of which is for development. Human and financial resources for the system are shown in Table 4.8.

MAFNR does not do research on sugarcane breeding, agronomy, and postharvest technology. It is, however, the principal actor in livestock and fisheries research and provides research services on plant protection in response to production problems. It has also been the major implementing agent of the national policy for food security and diversification. The ministry screens and tests varieties, cultivars, and technologies for potential high-value nontraditional exports, as well as for the many minor food crops that are locally important to the national food economy. By combining this screening and adaptive research with development activities, the ministry is able to perform a useful R&D function that other institutions could not perform as efficiently. Nonetheless, the overall picture does indicate that resources are spread very thinly across the ministry's research scope.

The need for improved procedures to focus, monitor, and manage this extensive R&D operation has set in motion a series of evaluations and proposals for reorganization. Research is currently organized along disciplinary lines. Whether this is the most efficient way to organize the ministry's research programs is still under discussion; however, a disciplinary focus does permit the ministry's research units to be flexible in the choice of commodities and topics to be covered. Where the generation of new technology is not the goal, but where making timely scientific inputs into production systems and monitoring new constraints are important concerns, organizing research on disciplinary lines may not be disadvantageous.

The University of Mauritius: The university is made up of four schools: the School of Agriculture; The School of Law, Management and Social Studies; the School of Engineering; and the School of Science. The extent of the university's contribution to agricultural research varies and is difficult to measure because the academic staff is only involved about 20% in research. However, research and teaching are not easy to disentangle, nor are they confined to any one school or department. Perhaps the key contribution of the university is its pool of scientifically trained human resources and its essential role in training future researchers, planners, and agriculturalists.

The participation of the university in agricultural research follows two streams. One is a long-term and basic stream in the area of policy research and natural resource studies where university-based research programs could have a clear identity. A second stream would focus on many areas where technology generation and transfer are concerned. Here, the university's role is best seen as short term and ad hoc; it provides opportunities through consultancies and other mechanisms that supply inputs for the research programs of the private sector and existing research institutions. Some formal recognition of these two streams of participation would enhance the university's contribution to the NARS in Mauritius.

The university is well placed to contribute in the areas of *policy research*, environmental studies, and information science. For example, university research can bring a multidisciplinary approach to the analysis of policy options for agriculture, science, and technological development. A notable example of such efforts is the "Mauritius 2000" study, which provides a generalized research and policy framework.[6] One important contribution to this has been made by the School of Agriculture, which conducts long-term studies in agroforestry and natural resource management in fragile environments. The Schools of Science and Engineering have the capacity to contribute in the bioprocessing of agricultural wastes and by-products, in biotechnology, and in information technology. All these areas will become increasingly important within the national scope of research in the years to come.

Where the generation of agricultural technologies is the goal, novel organizational mechanisms could be instituted to bring the university's scientific resources to bear more directly on development and production. Possible arrangements include closer links between the university and the private sector, e.g., through a university research and consultancy company, a joint partnership, or contracts. Formulas will need to be found to make the expertise of university staff available to industry without compromising essential training functions.

The Food and Agriculture Research Council (FARC) was established in 1985 to coordinate the research efforts of MSIRI, the technical services of the Ministry of Agriculture, and the university. It was to be responsible for (a) coordinating, monitoring, and promoting the research projects and programs of the different institutions engaged in agriculture, fisheries, and food, (b) ensuring as far as possible the proper dissemination and practical application of the results of any such research, and (c) advising the minister of agriculture generally on matters related to food and agricultural research.

FARC is composed of an executive chairperson and 12 members representing the ministry, research organizations, the University of Mauritius, various growers, and the food industry. It conducts its work through committees. The executive chairperson and two to three seconded personnel have responsibility for overlooking its tasks. The increasing complexity and importance of the policy and coordination function within agricultural research have placed greater demands on FARC. It has done much to promote collaborative arrangements between the academic, private, governmental and parastatal research components in the country.

FARC is funded by the Sugar Authority and the Ministry of Agriculture, Fisheries and Natural Resources with a present allocation of Rs 10 million (US$ 639,000). FARC occupies part of a building built with UNDP funds, at Réduit, to house the Regional Sugarcane Training Centre for Africa (RSTCA), with which it also shares a library. It maintains facilities for multiplying micropropagated plants, including tissue culture laboratories, greenhouse, and a multiplication area with humidity and shading control.

As FARC becomes more involved in the area of research implementation, particularly in the application of biotechnology, there may be a conflict between its policy and coordination functions on the one hand and its implementation functions on the other. There are ample reasons from an economic perspective why a small country like Mauritius should centralize its more expensive and sophisticated research facilities (such as those needed in biotechnology). On the other hand, one would not want to see FARC develop into one more institution implementing research, with the potential claim of being *primus inter pares* at the expense of its neutral and vital coordination functions and its linkages to policy and external actors.

Research in the private sector: The sugar industry is of course the major private-sector actor in research. Aside from its established role supporting the sugar research parastatal, MSIRI, individual companies also conduct experiments on their own. Advances in the physical and chemical composition of cane juices (Icery), obtaining cane seedlings from seeds (Perromat), correlation studies between cane yields and weather parameters (Walters), and sugar technology studies (Staub, d'Hotman, and Berenger) are a few examples of these individual efforts. On the other hand, firms such as Mon Desert Alma (sugar manufacturing), Ebène (cane breeding), and the Centre Agronomique du Nord (crop agronomy) have also made major research advances.

Over the years, many developments have been made by private companies in horticulture and animal production as well. At present, private firms are most active in research in the areas of aquaculture, horticulture (flowers and ornamentals), animal production, and food processing (fruit and vegetable canning, tuna canning, poultry, meat packing, making flour, and mixing feeds). Where possible, the private sector prefers to purchase research results directly, either in the form of overseas consultancies or as new technologies. They budget for the scientific expertise needed to scan and import technological innovations in their respective industries. It is largely in crop agriculture and aquaculture that private companies invest in more permanent facilities.

Public- and private-sector research partnerships and linkages are now being promoted and managed under the FARC umbrella. This has led to greater involvement of private-sector interests in national research planning and increases the opportunities for increased private investment in national agricultural research.

OVERVIEW OF THE NATIONAL SCOPE OF RESEARCH

In Table 4.9 it can be seen that the NARS has a rather broad scope. The broadest is under the ministry's research and development services. However, this broad scope is compensated by the downstream focus of research within the ministry, which primarily involves screening, testing, and adaptive research with close ties to extension and development. In terms of available scientific resources, the complex list of topics in the area of natural resource management has not yet been sufficiently staffed to develop the research programs and linkages needed to monitor and provide information for policymakers and resource users. Support for this type of research cannot come solely from the agricultural industry as in the past. Greater demands for conservation and optimal use of resources require a more diversified and complex research system.

Throughout its century of existence, the agricultural research system in Mauritius, in whatever form or shape, has mainly, but not exclusively, been involved with sugar. This commodity has funded the development of the system. In the future, the science base that was built for sugar will be increasingly expected to cover new topics and themes, which include biotechnology, food science and postharvest technology, mathematical modelling and system dynamics, mechanization, biomass processing, applied genetics, physiology and ecology, soil and environmental sciences, biological and chemical protection, agroforestry, aquaculture and energy farming, as well as animal and crop production systems, conventional or novel, large or small in scale.

While sugar will retain its central place, it cannot be expected to either fund or claim the lion's share of national research and training as it has in the past. The growing relative contribution of industry and tourism to the gross national product has created a growing tax base outside of agriculture. The prospect of greater intersectoral linkages between agriculture, industry, tourism, and the environment suggests that a diversification of the funding base to support it may be a real possibility.

As the links between agricultural development and other economic sectors grow stronger, agricultural research will increasingly be connected to the overall scientific research capacity in Mauritius, in industry, biology, engineering, and communications. This should enable Mauritius to continue its tradition of scientific excellence within small institutions.

Table 4.9. Mauritius: Research Components and Research Scope, 1992

NARS Institutions	Global Staples	Traditional Export Crops	Minor Food Crops	High-Input, Nontrad. Export Crops	Livestock	Soc. & Rural Engineering	NRM
MSIRI	Beans Maize Groundnuts	Sugarcane				Sugarcane processing Biogas Sugarcane by-products Potato processing & storage Drip irrigation	Land indexing
MAFNR Agricultural Services Forestry Services Fisheries Services	Beans Soybeans Rice Wheat	Cotton Tea Tobacco	Sweet potatoes Beetroot Cauliflower Eggplant Garlic Ginger Onions Tomatoes Okra	Citrus Mushrooms Turmeric Anthuriums Asparagus Cardamom Litchi Orchids Papaya Pepper Strawberries Mangoes	Animal production Swine Goats Sheep Ducks Beef cattle Dairy cattle Fodder/pastures	Groundnut storage Substrates Feeds Marketing studies Production systems Biogas Solar driers	Forest ecosystems Forest plantations Wild life conserv. Plant genetic resource cons. Soil conservation Watershed studies IPM Land indexing
University of Mauritius School of Agriculture School of Sciences						Farming studies Postharvest technologies Sugar technology Biogas	Agroforestry Range management Land-use studies

FARC: Provides overall coordination and facilities in biotechnology and tissue culture for research consortia and high-value, nontraditional exports.

SIERRA LEONE: USING LINKAGES TO COPE WITH SCARCITY AND TO CREATE STABILITY

Sierra Leone has a distinguished tradition of agricultural research and education, despite recent instability in public-sector institutions. The case study (Dahniya 1993) shows how external linkages and moving research closer to the farmer have been important factors in the continuing productivity of Sierra Leone's research system in this difficult political environment.

The country's varied production systems and commodities cover a large spectrum of the potential research scope, ranging from major global staples to traditional and minor food crops (orphan commodities) that are very important in the national diet.[7] Livestock systems and traditional exports are also important to the economy. Increasingly, developing nontraditional high-value export commodities is being considered, and exploitation of fisheries resources now provides a major source of national income. Sierra Leone also has one of the last remaining expanses of native forests on the upper Guinea coast, which are increasingly threatened by direct exploitation and agricultural activities. A forestry research action plan has been prepared, but much

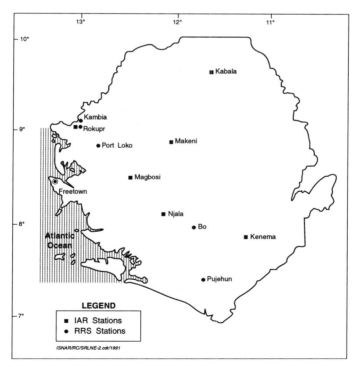

Figure 4.6. Sierra Leone and the location of its research stations

work is still needed to strengthen research on natural resource management, including fisheries, forestry, and land use. Given this broad and, in many cases, urgent demand for research, the support that can be provided by Sierra Leone's small research institutions requires careful consideration.

THE NATIONAL RESEARCH PORTFOLIO

Sierra Leone's agricultural research institutions are under the Ministry of Agriculture, Natural Resources and Forestry and are largely funded from the public purse.[8] Figure 4.6 gives a breakdown of expenditures by institution (see also Table 4.10). It is clear that rice research has been the leading component within the national research portfolio. The Rice Research Station (RRS) has released over 50 improved varieties and has identified effective methods for the management of mangrove and inland valley swamps. Rokupr was also the site of WARDA's research program for mangrove rice, which was carried out by national program staff and which made significant contributions to regional research (WARDA 1988). It was comparatively recently that work on upland rice varieties started, even though most of the rice grown in Sierra Leone is in the uplands. RRS is now also working on other small grains such as millet and sorghum.

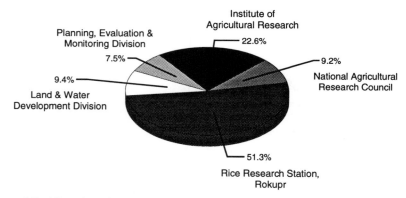

Figure 4.7. Allocation of government resources per institute, 1990

The Institute of Agricultural Research (IAR) conducts research on cassava, sweet potatoes, maize, groundnuts, and cowpeas as well as on alley farming and gender issues related to farming. Livestock, fisheries, and forestry research is presently at a low level, while research in the social sciences is fragmented and uncoordinated and has been severely limited by the mobility of local staff looking for more financially rewarding job opportunities.

Table 4.11 gives an overview of the national research portfolio, from which several conclusions can be made. First, the global staples and minor food crops of importance to Sierra Leone are well covered; research is making effective use of inputs from

Table 4.10. Research Institutions in Sierra Leone: Human and Financial Resources, 1991

Institutions	Researchers				Expenditures
	Total	PhD	MSc	BSc	
National Agricultural Research Coordinating Council, Ministry of Agriculture, Forestry & Fisheries (MANRF)	–	2	–	–	US$ 62,800 (Le 12,500,000) admin. costs
Rokupr Rice Research Station (MANRF)	21	4	13	4	US$ 376,900 (Le 75,000,000) 80% salaries
Institute of Agricultural Research, Njala (MANRF)	24	1	13	10	US$ 331,700 (Le 66,000,000) 76% salaries
Institute of Marine Biology and Oceanography, University of Sierra Leone, Fourah Bay	–	4	1	–	–
Other Contributors					
Faculty of Agriculture and Environmental Sciences, Njala University College	30[a]	13	16	1	–
Planning, Evaluation and Monitoring Division (MANRF)	NA	–	–	–	US$ 50,200 (Le 10,000,000) 86% salaries
Forestry Division (MANRF)	2	–	–	2	–
Land and Water Development Division (MANRF)	19	–	14	5	US$ 115,600 (Le 23,000,000) 74% salaries
Fisheries Division (MANRF)	28	–	–	–	–
Total	169	24	57	22	US$ 937,200

Note: Exchange is based on Le 199/US$ for 1991 (World Bank 1995).
a. Twenty-five percent of researchers' time is spent on research.

international centers. But there is a gap in the coverage of traditional exports and livestock. A question that arises is whether these can be covered by the existing national structures or whether linkages and new institutional arrangements are needed.

The gap in livestock research is difficult to justify on several grounds. Livestock is an essential part of the local farming systems and there are breeds and production systems for which research should be done locally. There is potential to make dramatic increases in productivity in the local breeds of *ndama* cattle, which are trypano-tolerant and fit well into low-input production systems. Links with the International Trypanosomiasis Centre in The Gambia should be strengthened to identify technologies that could be adapted locally. Production of small ruminants, swine, and poultry are widespread activities that are increasing in commercial importance.

Sierra Leone has been justifiably cautious about conducting research to fill the gap in traditional export crops. There is already a great deal of research being done in West Africa and elsewhere on coffee, cocoa, and oil palms. Partnerships and linkages to cocoa, coffee, and oil palm research institutes in the region and internationally would enable Sierra Leone to obtain information and some technologies without

Table 4.11. Sierra Leone: Research Components and Research Scope, 1991

Institutions	National Research Portfolio				
	Global Staples	Minor Food Crops	Livestock	Soc. & Rural Engineering	NRM
NARS Institutions					
Rokupr Rice Research Station	Rice	Findo (*Digitaria exilis*) Millet		Rice-based farming systems research	Management of riverine & swamp ecologies, grasslands, hydromorphic soils
Institute of Agricultural Research, Njala	Cassava Cowpeas Groundnuts Maize Potatoes Sorghum	Sweet potatoes Yams	Pasture improvement (limited research)	Farm structures Postharvest studies Farming systems Socioecomic studies	Soil management (agroforestry)
Other institutions					
Institute of Marine Biology and Oceanography Fourah Bay College University of Sierra Leone					Marine fisheries Oceanographic research
Faculty of Agriculture, Njala University College University of Sierra Leone				Farming systems Postharvest & storage Socioeconomics Marketing of staples	Plant genetic resources: identification and evaluation of local crop varieties, indigenous plants & trees Plant pathology
Land and Water Development Division (MANRF)					Data collection and analysis on soil & water utilization, including climatology & hydrological studies
Planning, Evaluation, Monitoring Services Division (MANRF)				Research impact Evaluation Marketing research	

Note: Sierra Leone's research scope does not cover export crops, either traditional or high-input, nontraditional.

major investments of time and facilities. The country may, however, be expected to pay for some of these technologies. But the extent of national research participation would be to include some traditional export crops as part of research in order to increase the profitability of farmers' production systems. This is still more realistic than establishing full-fledged commodity programs or institutes.

The gap in coverage for the high-value, nontraditional exports is justified given the poor transport and marketing infrastructure, and some technology testing on nontraditional exports such as shrimp has been adequately covered by the private sector.

Increasing the research capacity in natural resource management is an important concern. For example, there is a growing need for some research-based monitoring of the impact of the fisheries and aquaculture industries to determine sustainable levels of exploitation. With additional support, university-based institutes and programs, such as the Institute of Marine Biology and Oceanography (IMBO), could be mobilized. IMBO could play a larger role in supporting the regulatory function of the Fisheries Division by providing better information on coastal and marine resources. The economic importance of fisheries in Sierra Leone would certainly justify greater investment. In the management of land and water resources, the work of the Land and Water Development Division could be better integrated with agricultural research.

In Table 4.11, socioeconomics and rural engineering appear to be well covered; however, this is deceptive. Socioeconomic research is sporadic and not well integrated into other R&D efforts. Improved coordination, more efficient linkages to external sources of knowledge, technology, and funding, and closer links to producers should enable Sierra Leone to cover some crucial domains in its national research portfolio without creating new programs or institutions.

Changing policy environment and coordination: The basic objective of the agricultural development policy of Sierra Leone has been the attainment of self-sufficiency in food production, especially rice, and optimizing the production of export crops. Implementation of these general policies has fluctuated greatly. Beginning in 1972, the policy was based on the concept of an integrated agricultural development project (IADP). The IADP approach failed because of inappropriate government structures, inadequate government budgetary support, and lack of appropriate technological packages. In 1986, the Green Revolution Program (GRP) was formulated with the aim of accelerating the drive towards food self-sufficiency and economic recovery. This program also failed, primarily because of inadequate financial support. At present, there is no single overarching policy scheme to improve agricultural productivity and manage the resource base.

The recent revitalization of the National Agricultural Research Coordinating Council within the Ministry of Agriculture and National Resources augers well for two reasons. First, it will bring the various research components of crops, livestock, and natural resource management, including forestry, under a single umbrella and a common policy, which will promote an integrated systems approach to agricultural research. Second, it will provide uniform and improved conditions of service for those researchers under its umbrella, thus creating a scheme of service that is suited to the needs of both scientific research and rural development. Whether the coordination of the various research components will lead to consolidation into fewer institutions is not yet clear, nor is it certain if this would improve the efficiency and relevance of research. More important, the management of institutional and scientific linkages as well as the linking of research programs to farmer production systems is crucial to the continuing productivity of Sierra Leone's research institutions.

LINKAGES FOR INSTITUTIONAL SURVIVAL IN CONSTRAINED AND UNSTABLE ENVIRONMENTS

Political instability is a major policy issue for many African research systems. The Sierra Leone case provides a valuable lesson on coping with the political instability of the last 10 years. Like many other countries, Sierra Leone has been moving from more centralized and often authoritarian regimes to more pluralistic and democratic governments. The process is subject to many reversals and much turmoil. Whatever the merits of such a transformation, it is not easy to fund and manage research in times of rapid political change and uncertainty. Sierra Leone's NARS has been adversely affected by this process, but more remarkable has been the resilience and survival of the system, even in periods when donor support and representation was withheld or reduced.

In the turbulent policy environment of Sierra Leone, linkages to farmers, universities, and world science have been the lifelines of the system. They have enabled NARS personnel to maintain their scientific standing as researchers. Uncertainty, lack of resources, and a weak extension system has obliged the NARS to make more use of farmers' resources, knowledge, and experimentation (Monde and Jusu 1993). In many cases, this situation has even provided an opportunity to improve the relevance of researchers' work. The pressure in many commodity research programs to understand the production systems and farmer and consumer preferences has made it possible to select with greater precision those technologies for testing and adaptation that could be integrated into existing production systems (Dahniya 1992).

There were several examples in the Sierra Leone case where technology scanning, testing, and adaptation required a sophisticated scientific capacity to find those technologies with the greatest promise for the country's agricultural sector. As researchers moved closer to the farmer and assumed more responsibility for technology testing and transfer, it was important that the system's links to world science be maintained and reinforced. One of the reasons the NARS has been able to work effectively at farm level is because of the comparatively high number of well-qualified scientific staff. The high standard of qualification demanded for researchers may appear stringent, particularly since researchers are spending a greater portion of their time working directly with farmers, but it takes sophisticated scientific thinking to be able to generate information and recommendations that improve farming practices under low-input conditions.

Given the overwhelming problems of funding, infrastructure, and public service in general, one may well ask what it is that keeps the NARS functioning in Sierra Leone. One important factor is the close link between the agricultural research institutes and the university. This allows the NARS to tap into the scientific expertise of the faculty and to gain easier access to external information. This close university link has also served as a model for creating a status and management structure under the National Agricultural Research Coordinating Council that provides working and career conditions similar to those at the university. The university link and the opportunity to participate in regional research programs have allowed national researchers to maintain their scientific standing and credibility.

Another factor has been cooperation between the NARS and the international agricultural research centers. The programs of the centers have been essential for obtaining operating funds and overcoming the isolation of Sierre Leone's NARS. In some areas, such as mangrove rice research, Sierra Leone has been a contributor to the international centers rather than a receiver of their technologies.

In summary, the links to the regional and global science communities and the links with farmers are two key mechanisms that have enabled Sierra Leone's research system to operate successfully in times of crisis. Sierra Leone provides a good example of a NARS that acts as an effective linkage mechanism between the farmer and world science. (See Gilbert and Sompo-Ceesay [1990], which contains a description of the NARS as a model for linkage mechanisms.)

TOGO: FROM BALKANIZATION TO NATIONAL COORDINATION

The existence of many separate research organizations in a country is not always a good sign. In the case of Togo, the multiplicity of research institutions that existed prior to reorganization in 1992 was described as the "Balkanization" of the NARS (Aithnard and Gninofou 1992). The Togolese research system was made up of several government research institutes that functioned independently, even when they were funded by and accountable to the same sources, operated in the same agroecological zones, and served the same clients—small farmers.

When the case study was begun in 1990, Togo was a clear example of a situation where the multiplicity of institutions did not make sense, either as a strategy to diversify the sources of support or in order to be closer to the clients. The proliferation of institutes, each with its own status and policies, was not in response to national research policy but rather the result of the strong influence that external donors and collaborating agencies have exercised on the structure and direction of the research system. The low levels of productivity that resulted from this complex structure have sent a clear message to policymakers: develop national policies that will bring coherence to the national research system and make it more manageable.

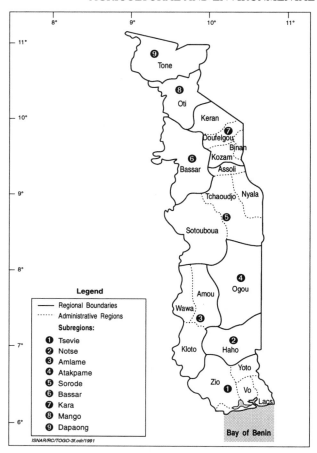

Figure 4.8. Togo

At the time of the study, Togolese research on food crops was done by four separate institutes, livestock research was done by yet a fifth, and research on traditional exports was done by CIRAD-IRCC's Togolese research program. Program formulation and policies, and even the status and remuneration of national researchers differed from institute to institute. While farmers tried to integrate food and export crops into their farming systems, the response of research to their needs was fragmented, costly, and ineffective. The total research effort was significant and in some cases of high quality, but the output was less than the sum of its parts.

The Balkanization of the system was in part the result of external institutions appropriating domains in Togo's national research portfolio before national policy had a chance to establish the mechanisms to plan and link these domains. Livestock research emerged out of the work of a German GTZ donor project. Research on natural resources such as soils, factor-based research on plant protection, and integrated pest management were planned within separate institutes built on the French colonial model, which had more prestige within the research system than the agronomic research institutes. Because of this, important links with universities among the units were difficult to manage, and many were underutilized, if they existed at all.

Policies for each institution as well as for each subsector were formulated independently. Some research policies were the product of bilateral agreements, as was the case with coffee and cocoa, and there were few opportunities to derive economies of scope because each institute protected its own domain in the national portfolio. The key objective for Togo was to bring the national system under one national policy. The key to coordination was the establishment of a national directorate for national agricultural research, which was given the mandate to develop research policies and management procedures to bring coherence to the national system as a whole.

The lesson from Togo is that multiple research structures in and of themselves do not lead to greater research capacity. Togo's pre-1992 system (Box 4.3) was less effective than it should have been, given the total resources it had available, because of its inability to channel these resources to meet the integrated needs of farmers. The

---

Box 4.3. Diversity or Fragmentation: Is there a Difference?

What is the difference between diversity and fragmentation? Honduras illustrates the case of diversity, where different types of research organizations orient themselves to work in a specific domain of the national research portfolio. This diverse set of organizations is able to tap a wider range of sources for funding, knowledge, and expertise. And they are often able to develop closer links to their client groups. This is what we refer to as an *institutional comparative advantage*. This type of system, built on institutional diversity, has the potential to be both effective and efficient in its use of resources. It does, however, create a strong demand for greater coordination at the policy level to orient and monitor activities.

Before 1991, the research system in Togo had at least as many research organizations as Honduras. However, the pre-1991 Togolese NARS was not marked by institutional diversity; rather, it was typified by a "Balkanization" or fragmentation of its research services. The various organizations that existed were all government institutes or departments, serving the same clients (Togo's small farmers) and drawing from the same pool of resources, government budgets, and donor technical assistance. The result was more than 12 government institutions with mandates for research on agriculture and natural resources divided among five ministries. In addition, there were also French tropical research institutes based in Togo (IRCC, IRCT, and ORSTOM) which, under French jurisdiction, received their policy and management from CIRAD while mandated to operate as part of the Togolese NARS and to respond to Togo's development policies.

This pre-1991 situation grew out of the evolution of each donor project into a separate institute, with policies originating in separate ministries and, in the case of traditional export crops, in separate capitals. Here, consolidating the institutions was the *sine qua non* for effective policy and management of the NARS.

transaction costs of so many organizations tapping the same funding sources, of separate policy and planning procedures involving the same set of participants, and of research policies and priorities being made in separate ministries and, in the case of coffee and cocoa, in different capitals were too great.

In 1992, the Direction Nationale de la Recherche Agronomique (DNRA) of the Ministry of Rural Development was created as the central body to regroup, plan, and coordinate research on agriculture and natural resources (see Table 4.12). This was the first step in the systematic management of the whole national agricultural research portfolio.

Table 4.12. Research Components and Research Scope in Togo

			National Research Portfolio			
Institutions	Global Staples	Traditional Exp.Crops	Minor Food Crops	Livestock	Soc. & Rural Eng.	NRM
NARS Institutes						
Direction de la Nutrition et de la Technologie Alimentaire (DNTA)					Processing Postharvest Human nutrition	
Institut National des Sols (INS), Min. of RDE&T[a]						Soil fertility & manage.
Centre de Recherche et d'Elevage d'Avétonou (CREAT)				Livestock prod. syst. Nutrition/ feeding		
Direction de la Protection des Végétaux (DPV), Min. of RDE&T						Plant protect. Weeds/dis./ pest control & manage.
Institut des Plantes à Tubercules (INPT), Direction Générale du Dévelopmt Rural	Cassava		Yams Sweet potatoes Taro			
French Tropical Research Institutes						
Institut de Recherche du Coton et des Textiles Exotique (IRCT)		Cotton				
Institut de Recherche du Café et du Cacao (IRCC)		Coffee Cocoa				

Note: High-input, nontraditional export crops are not covered by Togo's research portfolio.
There is no input to the national research portfolio from the Direction Nationale de la Recherche Agronomique (DNRA) in the Ministry of Rural Development, Environment, & Tourism; Office de la Recherche Scientifique et Technique d'Outre-Mer (ORSTOM); Ecole Supérieure d'Agronomie; or Ecole de Science.
a. Min. of RDE&T is the Ministry of Rural Development, Environment, & Tourism.

Given the depth and traditions that each of the various institutes had developed and the fact that some, such as coffee and cocoa, were in a privileged position in terms of resources and compensation, it will be a difficult task to bring the entire system under a single policy. Several of the key natural resource and postharvest components are still outside the focus of agricultural research, per se. However, this move towards consolidating the public research effort under one national policy is the first step that must be made if the system is to make better use of external inputs and if it is to attract the involvement of new actors.

LESOTHO: FINDING A ROLE FOR THE VERY SMALL RESEARCH SERVICE

Agriculture in Lesotho is based mainly on traditional smallholder production, where land is held under a communal land-tenure system administered by chiefs. While safeguarding traditional rights to land, Lesotho's government would also like to create an efficient commercial farming sector that can compete in international markets as well as providing cheap food for urban consumers. Agricultural production is mostly for subsistence, with livestock products in the form of wool (mohair) being the main source of income. The Republic of South Africa exerts a strong influence on Lesotho. Its economy and recent political turmoil have been felt in every aspect of the country. Moreover, efforts to develop commercial and export commodities must compete directly with the more intensive production systems in the Republic of South Africa.

Most of Lesotho is rugged and mountainous and lies at elevations over 1800 meters. Control of soil erosion is a top priority for the country. Most of the agricultural production systems are constrained by the rapid depletion of the natural resource base; it would not be an exaggeration to say that Lesotho's major exports are soil and water to South Africa.

Lesotho's research system faces urgent problems, but given its limited size and severe structural constraints, it is unlikely that the national system will be able to address these problems. New policies and coordinating functions need to be established to provide greater guidance for ad hoc research and development activities in the country.

INSTITUTIONAL FRAMEWORK

The Agricultural Research Division (ARD) and the Research Section of the Forestry Division of the Ministry of Agriculture, Cooperatives and Marketing share responsibility for agricultural research in Lesotho. ARD has one main station (Maseru Research Station), four branch stations (of which one is still in the planning stage), and seven substations (Figure 4.9). The number of researchers in the national system has varied between 17 and 25 between 1988 and 1992 (Table 4.13), with four researchers engaged in largely administrative duties.

Figure 4.9. Lesotho

Lesotho's NARS is undergoing a major reorganization to integrate research with training. ARD and the Lesotho Agricultural College will be brought together into a single research and training institution as part of the recently formed Faculty of Agriculture of the University of Lesotho. This arrangement will broaden the mandate and scope of the NARS; it will also raise a number of management issues to ensure that research remains oriented to the needs of the clients.

Table 4.13. Research Institutions in Lesotho: Human and Financial Resources, 1991

Institutions	Researchers				Expenditures[a]
	Total	PhD	MSc	BSc	
Agricultural Research Division, Ministry of Agriculture, Cooperatives & Marketing[b]	25	5	7	13	US$ 342,244 (maloti)
Other contributors					
Faculty of Agriculture, University of Lesotho	*NA*	–	–	–	NA
Lesotho Agricultural College[a]	26	2	13	11	NA
Total	*51*	7	20	24	–

Note: Exchange is based on malotis/US$.
a.   Figures are from 1988.
b.   Figures are from 1991.

Most research in Lesotho is actually conducted in development projects and by NGOs. One very important question involves establishing links to policy in order to provide guidance to and derive greater benefit from these activities. For the first time, a comprehensive survey of the agricultural research activities taking place within these projects has been carried out (Okello and Namane 1992). It documents the role that these projects play in agricultural research in the country, which is particularly relevant in research on natural resource management and in technology testing and transfer.

RESEARCH PORTFOLIO

Lesotho's policy on agricultural production aims at self-sufficiency in basic grains and the expansion of production in fruits and vegetables. Of the various crops grown, basic food grains (pulses and cereals) and livestock receive the most attention because of their direct contribution to domestic consumption (Peshoane 1988). Of equal importance is the development of agricultural production systems that conserve the natural resource base of this mountainous country. Second priority is placed on vegetables for domestic consumption, then on export crops, such as asparagus.

ARD research programs have neither been organized nor funded to reflect these priorities (ISNAR 1989a). For example, the conservation and management of soil and water, an overriding concern for the agricultural sector, has not been fully integrated into research. The university-based approach may enhance the ability of research to use and incorporate land-use studies and resource information systems, but it is not yet clear how the policy aspects of research on natural resource management will be handled under the new structure.

ARD conducts testing and adaptive research on global staples, principally beans, cowpeas, maize, sorghum, and groundnuts, and works with SACCAR-managed regional networks on these commodities. It also does research on small-ruminant production systems as well as poultry and swine. Wool and mohair are the Lesotho's main exports, but technology to meet the needs of this crucial sector comes directly from South Africa through the South African Wool and Mohair Marketing Board.

ARD's research on minor food crops covers a wide range of commodities, with several species of brassica, Swiss chard, carrots, garden peas, etc., under investigation (see Table 4.14). Activities are limited to testing new varieties and evaluating agronomic practices such as plant density, fertilizer levels, and time of planting, and no breeding work is involved. Some of the commodities in this category (e.g., asparagus) represent special niches that can be exploited only in very specific ways, and with external research inputs. What is most important is understanding the markets and the transport and handling considerations, but there is little formal work done on the socioeconomics of diversification and nontraditional exports.

Agricultural policy gives top priority to soil and water resource management and conservation. However, ARD's efforts in NRM are largely confined to pest management, the Range Management Program, and what remains of a donor-supported

Table 4.14. Lesotho: Research Components and Research Scope, 1991

			National Research Portfolio			
Institutions	Global Staples	Minor Food Crops	Livestock	High-Inp., Nontrad. Exp Crops	Soc.. & Rural Engin.	NRM
Agricultural Research Division	Beans Cowpeas Groundnuts Maize Sorghum Soya Wheat	Beetroot Cabbage Carrots Chick-peas Mustard (seed) Onions Peas Radish Tomatoes	Goats Nutrition, fodder Forage Poultry Sheep Swine	Sunflowers	Farming systems Storage, farm structures	Soils
Forestry Division, Min. of Agric., Cooperatives & Marketing						Silviculture Genetic re- source cons. Social forestry

Note: Traditional export crops are not covered by Lesotho's research portfolio.

farming-systems research project. Little formal research on forestry was conducted until the Lesotho Woodlot Project (LWP) recognized the need for local investigations. Out of this project, the Ministry of Agriculture created a forestry division with its own research section, which focuses on silviculture, forest inventory work, seed supply, and some sociological research. It employs two professional staff and three diploma-level foresters.

Natural Resource Management: Currently, government has assigned responsibility for natural resource and conservation research to development projects that have external support,[9] resulting in the division of this research among several departments and projects. This raises an important policy issue because research on the management of natural resources tends to be country and/or location specific and closely linked to policy. One of the major outputs of such research is to provide policymakers with the information they need to make the best use of land, water, and biological resources. Assigning the responsibility for research in this field to temporary institutional structures such as projects is expedient, on the one hand, because of the resources they make available and their close work with local communities. But, on the other hand, much of the useful information generated by these projects is either not used or lost. A small public-sector organization, such as ARD, may not be in a position to execute much NRM research but greater emphasis could be given to its coordination and monitoring functions.

Other aspects of NRM can be handled at the regional level, giving scope to a small country like Lesotho to conduct more formal long-term research. SACCAR's regional program has nominated Lesotho as a lead country in the region for soil and water conservation, and with this responsibility, ARD (and its successor in the university) has a unique opportunity to bring regional and international resources to bear on Lesotho's most important problem: soil erosion.

RELYING ON DEVELOPMENT PROJECTS AND REACHING THE POLICY LEVEL

Lesotho illustrates several lessons for countries with few formal institutions involved in research, where the greater share of agricultural and NRM research is carried out in development projects. The first lesson is that because research and development are often intertwined (Box 4.4), research institutions will have to take on a major role in guiding, monitoring, and using the work of development projects. They will not be able to see themselves purely as suppliers of technology for development.

Second, monitoring the research activities of development projects and NGOs will become increasingly important. In a small system, with only about 20 full-time researchers in crops, livestock, and forestry, there are many areas that cannot be covered within the scope of the national research institutions. In addition to monitoring the research activities in the country, these functions will include more responsibility for quarantine and for collating information on the introduction of technology into the country. For very small NARS, these nonexperimental functions may be unavoidable and may, in fact, be a positive feature in so far as they enable the NARS to guide the work of projects.

In such cases, an organizational structure that provides research with access to the policy level is crucial. Research in Lesotho has not had easy access to the policy level, in part because research is placed in the lower reaches of a ministerial bureaucracy. Research needs to be moved higher so that it can orient and, in some cases, regulate the R&D activities of development projects; otherwise, they may be at cross-purposes with each other or at variance with national policy objectives. There are cases where one NGO is planting tree seedlings on eroded slopes while another is introducing improved breeds of goats on the same hillsides. Technology also needs to be screened for its long-term impact and its suitability to local socioeconomic conditions. In Southern Africa, there is no shortage of technology that flows across the borders via official and unofficial channels, but the disparity in production systems between South Africa and its small neighbors is so great that this unregulated technology flow has led to wasteful and unfortunate applications.

Third, the NRM issue for research is not solely concerned with sustainable technologies for fragile or eroded lands. It is also concerned with decision making on the optimal uses of land, water, and biological resources. This is a role that small research systems need to pursue with great vigor. The new university-based structure in Lesotho could be useful in this regard if policy-linkage mechanisms are established.

Fourth, a small institution alone is vulnerable. It is often more difficult to retain staff in a small institution, and the departure of one or two scientists can cripple entire

---

**Box 4.4. The Link between Research and Development:
The Case of Guinea-Bissau**

In the past, research in Guinea-Bissau was done in a single organization, DEPA, within the Ministry of Rural Development. It had a modest infrastructure and limited human resources in terms of both numbers (fewer than 20 researchers) and level of training, and was entirely dependent upon external sources of funding. DEPA combined research activities with extension and development, a strategy that has several advantages for a small research system like Guinea-Bissau's. First, it ensures that the priorities of the researchers match the urgent needs of farmers. Second, it obliges researchers to work with extensionists and farmers. Third, this three-way dialogue establishes a common language for communication that facilitates the validation of research results by testing them in production situations. Fourth, it focuses research activities on adaptive research that can produce rapid results for transfer to producers, as well as facilitating the transfer and communication of information.

Since DEPA also executed development projects, it had a more realistic vision of the problems facing agricultural development, whether they were social, economic, technical, environmental, or policy related. All research activities began with an initial phase of rural development, and this led to the identification of those technical problems to which research could respond. Executing development projects also gave DEPA access to research funds that would not otherwise have been available.

While recognizing the advantages of DEPA's strategy, the execution of development projects tended to weaken the focus of the research programs. For example, research was forced to adopt more short-term perspectives and objectives. Also, researchers often spent too much time on development activities, to the detriment of their own research programs, and management costs were higher. DEPA spent a lot of time dealing with the administrative, technical, and financial management of development projects. At times, DEPA accepted development projects precisely in order to finance research activities, and as a result, DEPA frequently found itself orienting its work to meet the priorities of donors so that it would be doing the types of projects they were likely to support.

To improve the management and financing of research, in 1993, the government decreed the establishment of INPA, Instituto Nacional de Pesquisa Agricola, a public corporation belonging to the Ministry of Rural Development. The new structure provides clear financial autonomy and budgetary support for research, and the new management and personnel structure is more appropriate to the research functions of the institutions. However, greater attention will be needed to ensure that the link to development and the reliance upon close partnership with farmers and extension is not lost.

Source: Schwarz (1992).

---

programs. Linking to other science institutions in the country can be a way of providing stability and room for professional growth for researchers. This is the greatest promise that Lesotho's new university-based research structure provides.

The very small government agricultural research service will remain the core of the system. However, it should be structured in ways that promote close links to policy and that enable it to play a role in the management of the national agricultural technology system. The structure should facilitate access to external sources of knowledge and resources and allow for flexible management of external linkages. The smaller the research organizations, the greater the need for them to move higher in the policy and decision-making structures for agricultural research and development.

SUMMARY OF POLICY AND MANAGEMENT OF NATIONAL RESEARCH PORTFOLIOS

Our analysis of organizational structure and policy across seven countries reveals that public research organizations could play a greater role in formulating policies to manage the national research portfolio. In most countries, however, the organizational structures for research are ill suited to the policy and coordination functions that are most important for research systems in small countries. The tendency in such countries is to place research services at a low level within the national structure, making it difficult for them to manage their external linkages, liaise with policymakers and provide advice on options for agricultural development, and coordinate the development projects and NGOs that contribute to the national agricultural technology system.

Where there are several well-established institutions with distinctive sources of technology and funds, serving different sectors of the agricultural and natural resource community (as in Honduras and Mauritius, for example), a separate mechanism such as a national research council is an effective way to coordinate the research activities of each institution on the basis of its comparative advantage. Where public-sector research organizations have multiplied as a result of bureaucratic and administrative boundaries (Togo and Fiji, for example) some consolidation within the public sector is a good way to emphasize policy functions. This can, in turn, help to identify where the comparative advantages of government-based research lie and what policy can do to create an environment that promotes and guides participation by other institutions within and outside the country. Where public-sector organizations are very small and there are few prospects for major involvement by institutions outside the public sector or for direct financing by industry, then performing a policy advisory and linkage function between farmers and sources of technology (often outside the country) is one strategy that has proved successful in maintaining a coherent national research system.

The case studies highlighted other research policy issues as well:

- first, how to plan and integrate agricultural research and development within the larger environmental concerns of sustainable management of natural resources: land, water, and biodiversity;
- second, how research can contribute to diversifying the agricultural economy and income base of a country to keep agriculture competitive, both in world and domestic markets;

- third, how to take advantage of new developments in science and technology, such as biotechnology and information technology;
- fourth, how small countries can be effective partners in regional and international research as members of regional organizations and regional research systems and how they can make effective use of such networking to serve their national objectives, overcome isolation, and be recognized contributors to a global science of agricultural research for development.

These specific issues are dealt with in the next six chapters, which draw from the case studies and give examples from our study sample of 50 small countries.

Notes

1. See Pardey, Roseboom, and Anderson (1991), for a detailed quantitative analysis of research policy issues and global trends in the resources available to NARS.
2. It is important to remember that systems change. Indeed, Lesotho is in the process of moving research into the university and Honduras is in the process of "down-sizing" its public research services.
3. The national law that established DICTA was promulgated in mid-1992 after the study had been completed. DICTA's mandate is to encourage private-sector and producer involvement in research and technology transfer. The idea is for DICTA to promote greater centralization in evaluation and in the formulation of R&D policy while encouraging decentralized implementation of research and technology transfer by a diverse set of public, private, and parastatal institutions.
4. PROMECAFE, the Cooperative Program for the Protection and Modernization of Coffee Cultivation in Mexico, Central America, Panama, and the Dominican Republic, is supported and organized by IICA. The program seeks a common ground for cooperation in an industry where countries compete for the same export markets.
5. Until the mid-1980s, livestock and pasture research was in the research division and was conducted alongside crop research at the division's research stations. This support was mostly withdrawn when livestock and pasture research was moved to a separate division. Work in this area suffered greatly as a consequence.
6. The results of "Mauritius 2000" were presented (and favorably received) at a council meeting of the Association of Commonwealth Universities (ACU) held at the Réduit campus in April 1987.
7. Sierra Leone has an area of 72,000 km^2 and a population of 4 million people (1990 statistics), the majority of whom are members of rural farm households. Agriculture employs about 75% of the total work force. Shifting cultivation is the major farming practice, with a wide variety of crops grown under rainfed conditions.
8. Donor funding has not been a major component of the budget in the past five years, following the termination of the USAID-ACRE project (Dahniya 1993).
9. The conservation division, which has no research capability, has the overall responsibility for conservation work and, in theory, should be responsible for the projects as well. It is not clear whether the division is also responsible for research on natural resources.

5 Research on Natural Resource Management: A New Beginning or Just Another Agenda Item?

ENVIRONMENTAL ISSUES AND THE SCOPE OF AGRICULTURAL RESEARCH

Concerns about sustainability (see Box 5.1) have broadened the focus of agriculture, changing it from its traditional and relatively narrow production orientation to emphasize the maintenance of essential environmental processes such as regeneration and waste absorption.[1] These wider concerns pose new challenges to existing policy and research efforts and need new organizational approaches. In the past, the usual response of agricultural research organizations has been in one of two directions: they have either tried to include natural resource factors such as land, water, vegetation, forests, marine resources, and genetic resources in existing programs (thus broadening the criteria by which agricultural research is planned and evaluated),[2] or they have reorganized research. In this latter approach, some counties have taken steps to organize research around the natural resources themselves, which has meant reorganization into factor-based research programs or institutes, requiring some restructuring of institutions. Others have created multidisciplinary programs to deal with ecosystem research.

Research leaders in small countries have expressed concerns about the scope of research covering natural resource management and the environment: that it would stretch the limited capacity of existing organizations and move them into areas of

---

Box 5.1. Sustainable Agriculture

Growing concern over environmental degradation has led to a worldwide debate on environmental issues and the emergence of sustainability criteria for agricultural development (WCED 1987; Graham-Tomasi 1991; ISNAR 1993). Although sustainability is the "latest twist in the continuing elaboration of criteria by which agricultural development is defined and agricultural technology is evaluated" (Lynam and Herdt 1992: 205), there is a growing consensus on what it means. A more precise definition is given by Graham-Tomasi (1991: 81) as "the successful management of resources for agriculture to satisfy changing human needs while maintaining or enhancing the quality of the environment and conserving the natural resources".

long-term research where it takes longer to make an impact and where that impact is more difficult to demonstrate. For policymakers, faced with a decline in donor and government support for research institutions, creating new organizations to deal exclusively with natural resource and environmental problems in agriculture is also not an option. On the one hand, small countries cannot afford to expand the scope of research in order to make agriculture more sustainable, but on the other hand, can they afford *not* to address the NRM issues that affect the present and future productivity of their agricultural sectors? In this chapter, we argue that it is essential for small countries to address NRM concerns, and we propose institutional strategies for incorporating natural resource management into national agricultural research portfolios.

Natural resource management is particularly important for small countries because of the size of their agricultural sectors. For most of the small countries in the ISNAR study, agricultural production still constitutes a large portion of their gross domestic product, and the economic fate of these countries is closely related to the lasting productivity of their natural resource base. In addition, small countries are often characterized by rapid population growth, increasing urbanization, and a declining standard of living in rural areas. Linked to a deteriorating natural resource base, these trends are the result of economic policies that favor short-term production gains over long-term environmental costs. These problems are aggravated, in turn, by the many poor, small-scale farming units that adopt short-term strategies to enhance food security, along with technologies that result in short-term production gains.

Although environmental problems pose a threat to the economic future of developing countries, governments have been slow to react because of the complexity and cost of long-term environmental research. Expanding the scope of agricultural research to include natural resource management may not be feasible for most small countries if they are expected to work with their traditional research components alone. In essence, NRM research may require a fundamental change in the way that research gets carried out and in the number and kinds of organizations involved, but this change may carry unforeseen benefits for other aspects of agricultural research in small countries.

INTEGRATING NRM INTO AGRICULTURAL RESEARCH: RATIONALE AND POLICY IMPLICATIONS

One of the key issues in integrating NRM into agricultural research involves mobilizing a broader range of institutional actors. Working with NGOs allows research to move closer to the local level where the final decisions on resource use are made. And because NRM research is often more knowledge intensive than it is technology oriented, it requires institutions capable of gathering and storing information on the state of natural resources and trends in their use. For this reason, universities and specialized research institutes that conduct long-term studies are important contributors.

An NRM approach to agricultural research does not treat environmental issues and agricultural development as being incompatible or divorced.[3] On the contrary, in most developing countries, agriculture (including livestock, forestry, and fisheries) is the principal user of natural resources (Arntzen 1993). An NRM approach provides four reasons for NARS to incorporate natural resource management issues in their research agendas (Table 5.1). First, in situations where resources are underutilized, such incorporation may boost agricultural performance, especially of nonconventional plant products and certain forms of wildlife. Second, declines in agricultural production caused by a deteriorating resource base can be reduced or prevented by such an approach. Third, future agricultural development will increasingly rely on less productive environments and scarce natural resources; incorporating natural resource issues into research offers better chances of developing technologies and practices that are specifically suited to marginal environments and conditions. Fourth, the cost of rehabilitating degraded resources is usually much higher than the cost of preventing the degradation in the first place (Dregne, Kassas, and Rozanov 1992); timely incorporation of natural resource issues saves costs.

Table 5.1. Conditions and Benefits of Integrating NRM into Agricultural Research

Conditions	Benefits
Research that links NRM and agriculture	Agricultural research that integrates NRM issues can
• is multidisciplinary • has a cross-sectoral focus • cuts across natural resources • adopts a longer time horizon • incorporates interactions between different spatial levels	• boost agricultural development by developing underutilized natural resources, especially unconventional resources • reduce production losses due to environmental deterioration • offer better chances to develop relevant agricultural systems for the marginal environments that may have to be opened to agriculture in the near future • be less costly in the long run since prevention of degradation is usually cheaper than resource rehabilitation

Source: Arntzen (1993).

A second key issue in integrating NRM into agricultural research involves establishing and coordinating policy linkages. Policymakers need to be made aware of the importance of NRM research and its impact. Planning for agricultural growth without a broader environmental perspective can lead to a tendency to externalize negative environmental impacts in the following ways: (a) by transferring them to future generations, which, as mentioned above, can be costly because prevention is cheaper than environmental rehabilitation, (b) by exporting them, as is done with toxic waste, or (c) by shifting them to different sectors, such as developing valleys for irrigation projects at the expense of the livestock sector.[4] None of these solves any problems; they just postpone dealing with them.

Research on natural resources such as soil, water, vegetation, or animal populations is usually intended to increase, extend, or conserve their productivity. Sophisticated research on global environmental trends and processes is carried out by global organizations such as the International Union for the Conservation of Nature and Natural Resources (IUCN) and supported by global donor initiatives such as the United Nations Environmental Programme (UNEP) and the Global Environment Facility (World Bank/UNDP). But research on problems and processes in local situations is best done at the national level, where research on natural resource management involves the people who make the decisions: government, farmers, and local communities.

Some of the research on natural resources will have to be done at the regional level, or on a collaborative basis between countries. This is the case with shared resources, such as a river, a watershed, migrating game, or fish stocks. Research on salinization or iron toxicity in inland deltas in one country, for example, often provides useful information for other countries. However, a good understanding of a country's natural resource base is essential if the NARS is to be an efficient "borrower" of external knowledge and technologies. National scientific expertise, sufficient to provide advice in resource management issues, is an essential part of a small country's response to the demands of sustainable development in agriculture, natural resources, and the environment.

Research on natural resources requires long-term multidisciplinary analysis that cuts across different sectors and different natural resources and which incorporates linkages between the international, national, and local levels. This allows researchers to determine the optimal use of natural resources, the trade-offs and comparative environmental advantages of different sectors, and how environmental impact in one sector is passed to another. It also helps avoid some of the blind spots of sectoral planning in agriculture.

The cross-sectoral focus of NRM provides natural links between research and policy. When research on the environment incorporates institutional factors, people, and processes, it can have an important influence not only on market mechanisms but also in the development of new government programs and projects. The NRM approach allows researchers to take into account nonagricultural issues, such as population pressure, commercial demands, loss of arable land due to industrialization, as well as the differential influence of international, national, and local factors and the comparative environmental advantages and trade-offs of different subsectors, such as agriculture, livestock, and wildlife (Arntzen 1993).

Regardless of the scale of a NARS, natural resource management requires regular and sustained in-country research, not only because the long-term potential of natural resources tends to be site specific, but also because use and management systems vary greatly by ecological zone, the economic and policy context, and social strategy.

NRM RESEARCH AND THE INSTITUTIONAL PORTFOLIO

Research on the natural resources used in agriculture is long term; it is based heavily on information, field facts, and knowledge, as opposed to focusing on technology generation. In many of the small national systems studied by ISNAR, the potential exists to expand NRM research capacity by involving universities, NGOs, and the private sector.

Universities in several small countries are making important contributions to NRM research. The University of Mauritius, for example, has conducted a 20-year study on the performance of tree species used in agroforestry and their contribution to total biomass and soil conservation. In Botswana, many of the studies monitoring the state of rangeland resources and the impact of livestock have been conducted jointly by the University of Botswana's Department of Environmental Sciences and the National Institute of Development Research and Documentation. Honduras has based its forestry research within its universities. Also in Honduras, a private regional university, the Escuela Agrícola Panamericana, has assumed responsibility for research on biodiversity and the natural resource management aspects of integrated pest management. In Sierra Leone, the university's Institute of Marine Biology and Oceanography conducts research on marine resources.

In addition to the capacity to analyze links between sectors, national systems conducting research on natural resource management must also have the ability to access, store, and process information. The ability to gain rapid access to international sources of data is particularly important. Given their suitability for long-term information-based research, universities could have an important role to play in collecting, analyzing, and storing information on long-term trends in resource use and degradation. Global data-base networks are an important tool for factor-based NRM research; however, the ability to manage information and to cope with networks, methodologies, and new information sources is crucial (see Chapter 8 and Ballantyne 1993).

In many countries, both small and large, the primary organizations involved in collecting data on climate, water, and land resources have been land and water development divisions and government technical divisions (including aviation, transportation, and planning). While they have in many cases built up useful data bases, the organizations concerned have not had the ability to link the data to a wider range of resource users or to global data bases on environmental resources. Establishing links of this kind with universities and agricultural research organizations would be a cost-effective way of improving the usefulness of existing natural resource data bases.

In general, there is a need to link the various actors in this broad field more closely to national development and research policy. Without those links, useful information and important opportunities can be lost. NGOs are particularly active in the introduction of alternative land-use technologies in Sierra Leone and Lesotho, for example, but they lack access to both science and policy-making bodies. And they run the risk of introducing technologies that are counter to the needs of other sectors or resource users. This need not be the case, however. Botswana, for example, has a large and diversified NGO sector that conducts research on many environmental issues. NGOs have played a pioneering role in developing *veld* products (herbs and medicinal plants), which, in turn, has helped to diversify the agricultural sector. The development of new products in fragile environments can increase the motivation of local people to preserve those environments.

Care must be taken to avoid relying too heavily on the private sector for research aimed at development. The private sector does research only in areas where it has high stakes. It is more likely to introduce well-established commodities to new areas than it is to undertake new product development in resource-poor communities. However, these well-established commodities, if they are not resource-hungry mono-crops such as sugar, can contribute to the preservation of natural resources, either directly (through the use of tree crops and agroforestry) or indirectly (by increasing incomes). The probable short-term bias of private-sector research can be balanced by sound environmental policy and an improved capacity in public research institutions to monitor the resource base. These considerations should be explicitly reflected in national research strategies and plans, as well as in actual programs. Policies stimulating farmers and private enterprise to incorporate long-term considerations in their production strategies are needed.

There are also institutes in ministries of agriculture or rural development that focus on natural resource factors. This is particularly the case in francophone African countries, which have institutes devoted exclusively to research on and management of a specific natural resource, typically soils but also fisheries and forestry. Congo, for example, has a Centre National d'Etudes des Sols, Togo has an Institut National des Sols, and Mauritania has a Laboratoire National d'Analyse des Sols. In the past, such institutes tended to develop independent research programs, linking with other agricultural research institutions largely through the provision of services. In the future, as natural resource management is integrated more into the objectives of crop and livestock research programs, the opportunities for joint programs will be greater. These institutes are therefore an important resource for the future.

Forestry research is important in many small countries, but at present it tends to be isolated from research in other sectors, particularly crops and livestock. It is most often attached to separate departments for forestry development in ministries of natural resources, but it may also be found as a research component within a university, as in the case of the Centro Nacional de Investigación Forestal Aplicada (CENIFA) in Honduras. In countries where there are major pulp industries, the private sector maintains small forestry research units, as in Swaziland and Fiji. In franco-phone African countries, this research is conducted by parastatals, which primarily support the private lumber industry but also do some research on management issues. As the multiple use of forest resources becomes an increasingly important issue for developing countries, there will be (indeed, there already is) an urgent need to integrate forestry research with that on agriculture and other sectors, including wildlife conservation and tourism. Research on forestry has always presented problems because the objectives are so different: social forestry, commercial timber, tourism, or environment.

Despite the importance of marine fisheries for many small countries, programs in this area tend to be relatively new and weak. Nonetheless, fisheries research institutes now exist in most small coastal countries, and in three of these (Namibia, Mauritania, and the Seychelles) fisheries research is the largest single component of the national research system. In Sierra Leone, the university-based Institute of Marine Biology and Oceanography is responsible for research on the development and management of the country's marine resources. It has recently entered into collaborative research with the International Centre for Living Aquatic Resources Management (ICLARM). Fisheries research institutions in other countries are under ministries of agriculture, natural resources, and fisheries. These institutes are small, with only five to 30 researchers. Congo's Centre National de Pis`iculture de Djoumouna and Cape Verde's Instituto Nacional de Investigação Pescária are typical.

Only a few countries have research institutes organized around agroecological zones. Some examples of this are the Wau Ecology Institute in Papua New Guinea, the Desert Ecological Research Unit in Namibia, and the Atoll Research and Development Unit in Kiribati.

ENVIRONMENTAL CHARACTERIZATION AND RESEARCH STRATEGIES

No single strategy for NRM research is universally applicable. In determining a strategy for a given country, it is essential to begin by characterizing that country's agricultural and natural resource base. One factor that will influence the strategy adopted is biodiversity. For example, the diversity of crops, commodities, and ecosystems is lower in island countries, which tend to have less diverse natural, genetic, and production bases, usually as a result of their "insularism". However, the danger of extinction is greater for island species, which is a major reason for strengthening research on natural resource conservation in island countries. The degree to which agriculture causes pollution is also important. This is related to the degree of agricultural intensification and the use of inputs. With the exception of a few countries with intensive sugar, wood pulp, and timber industries, the negative environmental consequences of agricultural intensification are not yet great in our sample countries if the level of input use in agriculture is used as a measure. What may pose a greater environmental risk for small low-income countries (but is more difficult to measure) is the impact of extensive land-use technologies such as slash and burn and increased ranching and open grazing on fragile lands.

The relationship between agriculture and the environment affects any strategy for organizing research. The key parameters in this relationship are (a) the extent to which agricultural production is the main resource user in a country and (b) the population pressures on agricultural land that the country experiences (Runge 1992). Agriculture is the major user of natural resources in most small developing countries, but it is by no means the only one. An important starting point is therefore to consider the human population pressure on the agricultural resource base. Within our sample of countries, there are distinct differences in pressure on agricultural land. Three groups can be distinguished:

- *Group 1:* Over a third (38%) of the countries in our sample experience strong pressure on agricultural land. Most of them (14 out of 19) are island countries with very high rural population densities, averaging 540 persons per km^2. The remaining five (Rwanda, Burundi, The Gambia, Bhutan, and Laos) are small continental countries with somewhat lower rural population densities, averaging 280 persons per km^2.
- *Group 2:* Countries forming the second group are in the middle range of agricultural land pressure. They are generally continental states with small territories, such as El Salvador, Belize, Suriname, Honduras, Benin, Togo, Sierra Leone, and Guinea Bissau, but some of the larger island states, such as Fiji, and Trinidad and Tobago, are also included. All these countries have important crop and livestock sectors with legitimate demands for research services in addition to natural resource management.
- *Group 3:* The third group of countries currently experiences little pressure on their agricultural lands, but these lands are primarily extensive forests or permanent pastures, which require careful management to prevent irreversible degradation. The countries in this group include Guyana, Nicaragua, and Congo (which have large expanses of forests), Paraguay and the Central African Republic (with very diverse natural environments), and Mauritania, Vanuatu, Botswana, Mongolia, and Somalia (which consist mainly of arid and semiarid rangelands). The importance of agriculture in this group varies considerably.

ORGANIZATIONAL OPTIONS FOR NRM RESEARCH

As described above, our sample of small countries falls into three distinct groups, based on the relationships between agriculture and the natural resource base. This distinction, in turn, sets the stage for three broadly defined organizational strategies.

NRM PERSPECTIVE INTEGRATED INTO ALL ASPECTS OF AGRICULTURAL RESEARCH

The first strategy is for countries where agriculture already uses the bulk of natural resources and is the major cause of their depletion. In this classic Malthusian scenario, there are few resources to set aside and environmental buffers have already been exceeded. Population growth, along with growing poverty, places increasing pressure on resources as people are forced to choose between resource conservation and their immediate survival. For these countries, solving natural resource problems means improving agricultural productivity (see Table 5.2).

Table 5.2.   Strategic Approach #1: Natural Resource Management Perspective Integrated into All Aspects of Agricultural Research

Conditions	Research Strategy	Country Examples
• major share of land resources already under some system of production • high population density and land pressure • agriculture provides major share of national income and major source of employment	• continue research on commodity improvement for priority crops and livestock • organize research by agroecological zone or production system, ensuring a natural resource management focus • consolidate crop, livestock, and forestry research under coordinated planning for agriculture and natural resources	Rwanda, Burundi, Benin, Togo, Bhutan, Mauritius, Swaziland, Lesotho

Only through technologies that raise incomes and make production more efficient in its use of resources can environmental degradation be arrested. Efforts at resource conservation initiated by institutions outside the agricultural sector are likely to ignore the needs of agricultural producers and should not therefore be attempted. On the other hand, as resource degradation becomes a major constraint to increasing productivity and a major cause of rural poverty, research on agricultural production will need to consider land, water, and plant resources on a par with manufactured and human capital (Arntzen 1993). Reducing the costs and rate of resource depletion will be the key to increasing agricultural productivity.

In this first approach, crop and livestock production remains the major focus of research, but concerns about natural resource management urgently need to be integrated into ongoing research. This should be done within existing programs; research on natural resource management that is divorced from agriculture is not the way to arrest resource degradation.

To cater to these resource management perspectives, production-oriented research programs need to be organized and managed somewhat differently. A systems perspective and multidisciplinary research are both needed. Research should be planned, managed, and supported with longer time frames to allow the impact of production on the resource base to be measured. Finally, research-based solutions must be aimed at entire production systems and regions as well as at particular commodities.

RESEARCH ON CROP AND LIVESTOCK PRODUCTION WITH COORDINATED RESEARCH ON NRM

The second strategy is appropriate for small countries with important crop and livestock production sectors alongside large and relatively underused natural resources. Here, the demands of the established sectors claim continuing research support for production technologies. Research-based advice and information focused on natural resource issues such as pastures, forests, marine resources, water resources, and biodiversity, may be available from parallel but coordinated institutions. In these situations, a coordinated strategy based on institutional linkages between research on commodities and natural resources is desirable (Table 5.3).

Table 5.3. Strategic Approach #2: Research on Crop and Livestock Production with Coordinated Research on Natural Resource Management

Conditions	Research Strategy	Country Examples
• country will rely on national production to meet a significant share of food security needs from favored areas • important areas already devoted to traditional export crops • medium to low population density and land pressure, some resources under-utilized or undervalued	• continue NARS focus on production research for major commodities in higher potential areas • include environmental concerns in research policy and program formulation: some NRM research on marginal areas for equity reasons • strengthen linkages with resource management and environmental research institutes • provide policy guidance and information management mechanisms for NGOs and development projects doing some local-level NRM research	Honduras, Nicaragua, Panama, Belize, Jamaica, Trinidad and Tobago, Suriname, Guyana, Equatorial Guinea, São Tomé e Príncipe, Botswana, Congo, Central African Republic, Somalia, Papua New Guinea, Laos

Growing concerns with the environment should not lead to the neglect of research on increasing agricultural productivity. An important aspect of research on productivity is the search for more efficient and economical ways of using natural resources in the production process. Reduced levels of agrochemicals, increased productivity per unit of land, and more efficient use of water are some of the key ways that more efficient agriculture can contribute to the conservation and management of natural resources. In many cases, concentrating on increasing the efficiency of production in favorable areas may reduce the pressure to expand production into less favorable, more fragile areas.

Research in these areas is usually carried out by public-sector organizations with a commodity perspective; however, this research does not necessarily address the needs of fragile natural resources such as forests, permanent pastures, tidal areas, marine resources, and genetic resources. Research on these resources may need to involve other institutions in other subsectors, such as forestry or fisheries, as well as universities and NGOs. Managing linkages and formulating intersectoral research policies are particularly important.

NRM AS THE ORGANIZING PRINCIPLE OF THE NATIONAL AGRICULTURAL RESEARCH SYSTEM

The third strategy is for countries where agricultural production is low and precarious, largely because of a severely degraded or fragile resource base. In these countries, the main natural resources used in agricultural production (soil, water, and vegetation) are virtually exhausted or would be rapidly depleted if brought into production. In some countries with large territories, there are broad expanses of fragile lands or marine resources to manage. In these countries, where resource constraints and resource management issues dominate all other production constraints, agricultural research should be organized around key resources or agroecological zones. Commodity and production technologies can be introduced as a result of improved understanding and management of those natural resources (Table 5.4).

In these countries, introducing new and improved agricultural technologies requires careful study of resource limitations and potential. Research on the natural resource base is essential if opportunities for improved agricultural production are to be created and the precarious balance between production and conservation is to be maintained. Coral islands in the Pacific and severely eroded and desiccated islands surrounding Africa are typical examples. In one such island country, Cape Verde, research is focused on water harvesting, water management, and soil conservation (Sabino 1992). Identifying appropriate technologies for commodity production is both *derived from* and the *result of* research on the country's natural resources.

INTEGRATING NRM INTO NARS PORTFOLIOS: A STRATEGIC OPPORTUNITY FOR SMALL COUNTRIES

Research on NRM issues is more complex, involves a broader set of institutions, requires more complex linkages among local, regional, and global actors, and has longer time frames than does traditional commodity or disciplinary research. For this reason it appears to be a daunting challenge for small-country research systems that are already overstretched and have few opportunities to increase their size. Is NRM an extra item on the agenda of already overburdened NARS, or does it represent a new opportunity for small countries? We regard NRM as an opportunity to increase the returns to research investments:

- Policymakers rely on information from research to make difficult decisions on resource use. Research on NRM can provide this information, enabling governments to use their natural resources sensibly and in a sustainable manner.

Table 5.4. Strategic Approach #3: Natural Resource Management as the Organizing Principle of the NARS

Conditions	Research Strategy	Country Examples
• very small hectarage and output per agricultural commodity • extremely fragile or badly eroded resource base • major national income from a renewable natural resource: fisheries, forestry, open grazing	• research on agricultural production mainly involving testing and adapting external technologies • improved knowledge of productive potential and optimal use of resources for introducing new agricultural technologies • improved knowledge of farming systems to screen technologies for farmer preference and adoption • knowledge on the state and potential uses of natural resources acquired and stored for resource policy and decision making	• *Cape Verde:* fisheries, soil and water conservation and harvesting • *Mauritania:* fisheries, soil and water conservation, permanent pastures • *Namibia:* fisheries, soil and water conservation, permanent pastures • *Seychelles:* fisheries, genetic resources, soil conservation • *Mongolia:* pastures, soil conservation, genetic resources • *South Pacific Island Nations:* fisheries, genetic resources, soil conservation • *Eastern Caribbean States:* genetic resources, soil conservation

- Most small countries must screen external technologies to select those that can be adapted or introduced. Improved knowledge of the natural resource base is crucial to intelligent and effective borrowing.
- NRM research in small countries can provide direct inputs to local resource users. It can help guide the decisions of resource users on the most efficient and sustainable use of existing resources.

We found three organizational strategies for dealing with NRM research. Two involve merging NRM and production research perspectives into existing institutions. The third relies on linkages with other institutions, such as universities. Universities seem to have a comparative advantage in assessing the environmental impact of agricultural production and they are well placed to conduct long-term studies, large-scale resource surveys, and policy research on environmental issues. NGOs can also make a useful contribution, especially at the grass-roots level, although their activities are seldom formally linked to national policy objectives.

Whether a national strategy aims to integrate natural resource management perspectives into production research, to organize all agricultural research around natural resource factors and ecozones, or to establish linkages between agricultural research and other institutions, policies that cover both agriculture and the environment will be key components of the national effort. The objective is to link environmental and agricultural development policy, including proper mechanisms to fully incorporate natural resource management issues into agricultural policies (Box 5.2). In principle, these issues should be raised from both the environmental and the agricultural side.

Box 5.2. The Role of Research in Developing a Package of Policy Instruments for Sustainable Natural Resource Management in Agriculture

- to assess the role of agriculture in national development
- to identify nonagricultural factors that increase pressure on the natural resources that are vital to agriculture
- to develop measures and criteria for more and better monitoring of the natural resource base
- to identify optimal resource use based on local natural resources and socioeconomic conditions
- to apply a total-factor-productivity approach to promote efficient resource use within existing agricultural production systems
- to make available technological and production options for better utilization of natural resources
- to treat rural poverty as an issue of natural resource management as well as of equity
- to assign institutional responsibility for resource management and establish account-ability

---

To design policies in countries where NRM is a priority, decision makers have to consider many complex factors. The major ones are optimal resource use, trade-offs between sectors, the availability of environmentally sound technologies, government legislation on the environment, and the establishment of environmental standards. Given the complexity of these issues, an effective policy on natural resource use depends upon close linkages between researchers and policymakers. Once these are established, information on changes in the natural resource base can be channeled from researchers to policymakers and can contribute to formulating policy and assessing its impact. Such information for decision making becomes just as important an output of the research system as new technologies.

Notes

1. This formulation is from Arntzen (1993) who has analyzed the relationships between agriculture and the environment in the small countries of Southern Africa.
2. Crosson and Anderson (1993) have applied a total-factor-productivity approach to research on production, which incorporates sustainability concerns.
3. Nearly all the small countries in our sample have signed the Convention on Biological Diversity, which binds governments to develop strategies to integrate the conservation of biodiversity with their national development goals.
4. This section is based on the analysis of Arntzen (1993).

6 Diversification and High-Value, Nontraditional Exports: Can Research Open "Windows of Opportunity"?

THE GLOBAL POLICY AND ECONOMIC ENVIRONMENT

For many in the North, the epitome of the small country is a tropical paradise with coconut trees and banana, sugarcane, and coffee plantations. In actuality, when a country depends on one or two of these crops for most of its income, it is far from "paradise". The production of these crops dominates the national economy, which, in turn, is dependent upon the whims of consumers in distant parts of the globe and the latest dietary fads. In a country like Vanuatu, for example, coconuts were the only significant export for a long time, and the country's economy was severely affected when consumers stopped buying coconuts and coconut products.

Other countries have been in similar situations. The economies of Fiji, Mauritius, and Barbados have been dependent on sugar; in El Salvador, it has been coffee; and in Honduras and Saint Lucia, bananas. Cocoa in São Tomé, groundnuts in the Gambia, and cotton in Benin are other examples. Despite a gradual trend towards diversification over the past 30 years, in many of these "monocrop economies" a single crop is still cultivated on most of the available arable land, making a significant contribution to employment and export earnings (Table 6.1), but also making the economy vulnerable to what are sometimes extremely volatile markets.

Long-term trends for traditional export crops point to a world market characterized by increasing production and stagnant or, at best, slowly growing markets. To survive and remain competitive, countries must cut their production costs and achieve economies of scale. This is difficult for the small country, constrained both by the area it can cultivate and by the risks of overdependence on a single crop. An attractive option for such countries is to diversify into new, high-quality and specialty products for small market niches, such as exotic fruits and vegetables, spices and nuts, and ornamentals and other nonfood products. For these so-called nontraditional export commodities, small quantities can provide significant amounts of income, and all of them have scope for significant growth in world markets as incomes rise. Box 6.1 describes the full range of products covered by the term "high-value, nontraditional agricultural exports".

Most developing countries have recognized the need to diversify into high-value, nontraditional exports and have embraced this as a policy objective. This has also

Table 6.1. Dominant Agricultural Exports in Selected Small Countries, 1990

Country	Traditional export	Value of export commodity (US$)	Total value of agricultural exports (%)	Total land under export crop (ha)	(%)
Chad	Cotton	91,050,000	59	200,000	6.2
Gambia	Groundnuts	6,610,000	51	90,000	50.6
Benin	Cotton	91,650,000	83	116,000	6.2
São Tomé e Príncipe	Cocoa	3,300,000	89	30,000	81.1
Equatorial Guinea	Cocoa	7,500,000	89	74,000	32.2
Mauritius	Sugar	341,970,000	93	76,000	71.7
Swaziland	Sugar	162,900,000	78	40,000	24.4
Fiji	Sugar	185,3 50,000	85	68,000	28.3
Vanuatu	Coconuts	5,137,000	46	69,000	47.9
Western Samoa	Coconuts	4,370,000	47	42,200	34.6
Solomon Islands	Coconuts	4,832,000	29	38,000	66.7
Barbados	Sugar	31,740,000	55	11,000	33.3
Belize	Sugar	43,000,000	55	24,000	42.9
Dominica	Bananas	31,499,000	88	–	–
El Salvador	Coffee	267,390,000	83	174,000	23.7
Guyana	Sugar	81,850,000	79	40,000	8.1
Honduras	Bananas	367,800,000	55	20,000	1.1
Honduras	Coffee	183,900,000	28	150,000	8.3
St. Lucia	Bananas	73,962,000	87	5,346	29.7

Source: FAO (1992).

Box 6.1. What Are High-Value, Nontraditional Agricultural Exports?

The range of products covered by the concept of nontraditional high-value agricultural exports is broad. Some are new crops introduced to meet the demand for fresh produce in the industrialized countries. These include vegetables (such as broccoli, brussels sprouts, green beans, snow peas, asparagus) and melons, tropical fruits, and nuts (such as mangoes, starfruit, lychee, passion fruit, macadamia, papayas, and avocados).

Others are food crops that have long been part of the traditional food systems of some developing countries, and which are now finding their own specialty markets abroad, often in communities of immigrants from the country of origin. Included here are tropical foods such as fresh taro, cassava, okra, plantain, and breadfruit. Some of these, notably cassava, have considerable potential for a range of uses, including animal feed, allowing the producing country to increase its earnings through processing. Others have a more limited, but still lucrative, appeal: the traditional Andean food grain, quinoa, is now reaching specialty health-food markets. Another group consists of spices, stimulants, and essential oils, such as vanilla, ylang ylang, kava, jojoba, cardamom, allspice, black pepper, and pyrethrum.

Besides these agricultural exports, aquaculture products, principally shrimp, are growing in importance in many small developing countries. There is considerable potential for aquaculture technology to be applied to other shellfish as well.

been supported by donor and international agencies, which actively promoted diversification throughout the 1980s. The Caribbean Basin Initiative and its support for nontraditional-export projects provided preferential access to the US market for this long-disadvantaged region. The various Lomé Conventions have promoted nontraditional exports of fresh fruit and vegetables from ACP (Africa, Caribbean, and Pacific) countries to the European Community. While the main concerns have been access to markets and transport, there has also been some support for research and technical assistance in areas such as storage, packing, and transport techniques.

High-value, nontraditional exports have been touted as opening up "windows of opportunity" for small developing countries. But it is a much more complex situation than it appears on the surface, and experience suggests that nontraditional commodities are in fact a high-risk venture for these countries, from both a production and a marketing viewpoint. These windows to markets in the developed world open and close abruptly with the changing winds of economic fortune, trade policies, and consumer preferences. The markets themselves are highly competitive and demand the highest standards in terms of consistency of delivery and product quality (Durant and Blades 1990). Many are protected by various forms of trade barriers, including tariffs and duties. Phytosanitary and quarantine regulations must also be observed in both the producing and the importing countries.

Our study revealed a cautionary tale, where traditional types of research are not an effective way to approach these new commodities. If research is to respond to the demands arising from nontraditional commodities and markets, new partnerships and flexible structures are necessary. In this chapter, we examine how the research portfolio can be managed both to identify the opportunities for small countries and to reduce the risks they face in attempting to capitalize on those opportunities.

CAN A SMALL COUNTRY COMPETE?

While donor agencies and national policymakers have promoted nontraditional exports with enthusiasm, the prospects for small countries competing in these markets may have been overstated. Economists have argued that "small countries provide few opportunities for economic diversification or economies of scale, and scope barely exists for effective competition and operation of market forces" (Howlett 1985: 113).

Indeed, those developing countries with strong nontraditional export subsectors have so far tended to be middle-income countries, which have relatively strong internal markets for the products as well as experience with other commercial exports (Islam 1990). These internal markets provide an opportunity for new products to be tried out. They also act as a cushion when export markets are elusive, when they shrink, or when they are oversupplied. With their smaller internal markets, small countries are at a disadvantage in this respect.

The history of most nontraditional exports demonstrates dramatic shifts in the number and types of producers and the changing markets for these crops. Often it is the very success of a nontraditional export crop that results in its failure as a feasible product in many small countries. Sometimes production shifts to an industrialized country that was formerly an importer. Ginseng, for example, is now produced in western Canada. In other cases, small markets are quickly saturated when producers in several countries all discover the same opportunity at the same time, as happened with cardamom in Central America (see Box 6.2). In still other cases, such as the development of kiwi fruit in New Zealand, a good product is seized on by other countries that are more competitive, and they eventually take over much of the market initially developed by the small country.

Box 6.2. Cardamom in Central America: A Cautionary Tale

Cardamom prices were very high in the early 1980s, following production difficulties in the world's largest exporting country, India. As a result, several coffee-producing countries of Central America encouraged their farmers to grow the crop, with support from research institutes such as the Instituto Hondureño del Cafe (IHCAFE).

The only problem was the popularity of diversification as a policy: everybody was doing it, and doing it in precisely the same way. Given the small quantities of cardamom that enter the international market (only 10,000 tonnes a year; a quantity that can be produced from less than 30,000 hectares), even a single small country can have a marked effect on supplies. In addition, global demand was growing only sluggishly. As production from several Central American countries came onto the market at the same time, prices fell drastically. They are now so low that research investments have been all but lost. In Honduras, IHCAFE has given up promoting diversification into cardamom.

In these risky environments, market conditions and prospects should be analyzed before investments in commodity research and development are made.

Small countries can all too easily be squeezed out by larger countries with better marketing and transport infrastructures. Although small quantities of produce can generate significant income, there is no disputing the fact that small countries do not have a comparative advantage over larger producers, especially in the longer term.

DIVERSIFICATION AROUND TRADITIONAL EXPORTS

Diversification does not mean that a traditional crop will cease to be important. Often, it is easier to diversify into high-value, nontraditional exports when these are produced in association with, rather than as replacements for, a traditional export. Where high-value, nontraditional exports fit well alongside a traditional export crop, diversification can take place within and around the traditional export rather than away from it (Bonte-Friedheim 1992).

Countries with a traditional export crop have developed transport, postharvest, and marketing infrastructures and systems that can also be used for nontraditional exports.

In many cases the institutions involved in developing or marketing the traditional export crop are among the first to be associated with the new crops. They are well placed to seek crops that will be "good companions" for the traditional export crop, helping to break pest and disease cycles or to restore soil fertility, as well as to increase the income-earning opportunities of producers and to minimize labor peaks and troughs.

Diversification in Honduras, for example, is taking place both within and around its two traditional export crops, bananas and coffee (Wyeth 1989). It was the two banana multinationals, Standard Brands and Chiquita Brands, that began developing melons and grapefruit for export during the late 1970s. For other new commodities, such as cardamom, macadamia nuts, and pimento, the Coffee Institute took the lead (Contreras 1992; Zacarías 1992). The Central American and Caribbean banana-exporting countries had an advantage for diversifying their exports in the existence of export handling, transport, and marketing channels for fresh produce. In Honduras, as in other Central American and Caribbean countries, some of the new fruit and vegetable exports could occupy extra cargo space on the refrigerated banana boats. For other commodities, new marketing, handling, and transport systems had to be established, but the expertise acquired in the traditional export sector was available to support these initiatives.

THE ROLE OF RESEARCH

As we have seen, however, the promise of quick gains from new export crops is accompanied by high risk. The chief objective of research must be to reduce some of this risk, but this kind of research differs from research on staple food crops, livestock, and traditional exports. Research on high-value commodities, for example, emphasizes different research functions, with much more emphasis placed on information and market intelligence. Central to any discussion of research on nontraditional crops is the recognition that these differences entail quite different roles for the agencies involved in the research. Public-sector organizations will be required to act as catalysts and facilitators rather than researchers. And the critical issues of marketing, socioeconomics, and postharvest handling and technologies can be addressed in many cases through partnerships between the public sector and producers.

Carefully considered research investments by small countries can lead to success, and experience suggests that the keys to this success are diversification around traditional exports, gearing research investments to key functions, and timing these investments carefully in relation to the product development cycle. The clear allocation of institutional responsibilities in a sound partnership between the private and public sectors, backed by strong policy support, is also important. Provided these conditions are met, high-value, nontraditional exports can bring significant benefits to small countries.

KEY RESEARCH FUNCTIONS

For nontraditional exports, the traditional sequence of research for technology generation (Research Program > Technology > Producer > Consumer) is reversed, beginning instead with a focus on the consumer: Consumer > Producer > Technology > Research Program. In most cases the sequence moves only as far as the technology stage, i.e., the testing and evaluation of available technologies. Doing the research required to develop new technologies is usually a poor option for a small country, since this research normally bears fruit only after the market niche has been filled by others, or worse still, it may actually enable others to fill it.

For a small country, then, research on nontraditional, high-value exports is not concerned with solving long-term constraints but with short-term, near-market research, aimed usually at the introduction of new crops into existing production systems. The research is therefore concerned more with technology scanning and testing than with technology generation. This means that information and market intelligence will be vital, especially at the outset (see Box 6.3).

Box 6.3. Links to External Sources of Information

Through networks and other mechanisms, research organizations maintain many links to other sources of information and technology. However, most of these traditional links are not useful when new export commodities are introduced. Needed instead are links to specialized sources of technology in more-developed countries or in larger developing ones. These sources are generally private. For example, for research on anthuriums and orchids in Mauritius, the public agencies involved could not expect to receive information from their traditional partners; instead, they had to rely on private industry in the North.

Because of the competitiveness of the industry and the small size of the markets, these specialized sources may prove reluctant to share information and technology. Tapping them should, on the whole, be left to producers, who can order seed from recognized suppliers. Often too, it is producers and consumers who bring seeds across national borders in a spontaneous technology-transfer process, and they must be encouraged to share some of their new materials with private-sector and national research organizations for research purposes.

At the institutional level, technical study tours, visits by consultants, and industry and professional meetings are among the major ways in which information on nontraditional exports is exchanged. Small countries should encourage their scientists and development specialists to participate in relevant events. This can be a rather hit-and-miss approach to intelligence gathering, but it is better than nothing.

Public agencies can also build on these contacts by systematically storing and circulating the information picked up by individuals. In this way the information enters the knowledge base of the national research system as a whole, rather than remaining the property of one specific, and perhaps transient, part of it.

Once new export opportunities have been detected, postharvest research and marketing become the key considerations (Kaimowitz 1991; Rojas 1993; Zacarías 1992). Few national research systems have established strengths in this area, which is a vital one for countries remote from their markets, as small developing countries often are. A review of agricultural diversification programs in the South Pacific recommended an approach to marketing systems research similar to the more familiar farming systems research used for staple food commodities (Hardaker and Fleming 1990).

Another important function for the national research system is regulation, including quality control, which is an important consideration for markets demanding high and consistent quality. Phytosanitary regulation and quarantine are important when technology is introduced from other countries. Standards for organic produce may also need to be set and monitored.

The advisory function is also an important task for the public-sector core of the national research system. With pressures from senior ministers, donors, and industry to diversify, policymakers need sound advice based on scientific knowledge about the potential productivity of a commodity and its likely impact on the resource base. Links between producers and policymakers are especially important, given the volatile nature of nontraditional export markets. Effective policies can help cushion some of the risks faced by producers.

Coordination may be needed to avoid duplication of efforts. The number of short-term and ad hoc initiatives in this domain, while useful and more efficient than long-term research programs within large institutions, can pose problems. With limited resources to spend on risky commodities, a small country can ill afford to have more than one research initiative on the same commodity.

THE PRODUCT DEVELOPMENT CYCLE

The introduction and development of nontraditional export crops follows a clear product development cycle with four phases. In the first phase of the cycle, products are developed, normally through conventional technology generation. Markets are as yet nonexistent. Strategic and applied research are carried out to gain a better understanding of the product's biochemistry and physiology and to develop improved varieties through breeding. Thus, researchers in New Zealand took an obscure Chinese fruit, introduced it to New Zealand, and invested several years of research in its improvement. It became known as the kiwi fruit.

The second phase of product development begins once a product is seen to have high growth potential and when markets open up. There is a shift away from applied research towards adaptation to local conditions. Tried and tested technologies move rapidly into commercial production. At this stage, market knowledge is essential and is complemented by research on postharvest handling, transport, and consumer acceptance. Traditional agronomic research begins to play a role, especially in determining appropriate production inputs and practices suited to local conditions.

The third phase is characterized by market saturation. Once there are proven technologies and large markets are known to exist, more and more countries enter the market. As production increases, prices come down. The research system comes under increasing pressure to reduce unit costs and create economies of scale. As time goes on, plant protection problems arise and environmental concerns mount, increasing the need for a broader and more permanent research effort.

The fourth phase is one of market decline. Once market prices start to fall, producers tend to move their research into areas that will lower the costs of production and thereby keep the product competitive. For those that stay in, the nontraditional export becomes a "traditional" product: one on which research becomes well established, being organized along traditional commodity lines and requiring continuous and long-term support.

PHASING RESEARCH WITH THE PRODUCT CYCLE

Research leaders and policymakers can use the product development cycle to gauge the timing and level of their investments in research. Research activities and, hence, costs change as the cycle progresses. Figure 6.1 illustrates the inverse relationship between a product's profitability and the costs of the research required to develop and support it. Costs are high during the product development phase when markets are small. They fall rapidly as sales and turnover expand but rise again as conventional commodity research becomes necessary in the later phases.

The goal of the product development strategies adopted by many large companies is to get a steady stream of new products onto the market as the profitability of older products declines. This is an effective strategy when a company has the resources to invest lavishly in research and development, but for small developing countries lacking these resources, it is not feasible. A small country, then, is less likely to be

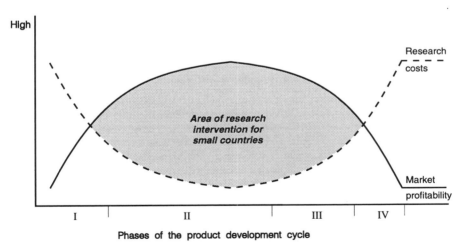

Figure 6.1. The product cycle and research and development costs

involved in the costlier early stages of the cycle. It can, however, time its entry onto the market to take advantage of recently developed technologies and products that require only testing and market research.

Thus, while phase I research is pursued by their larger or more developed competitors, small countries should emphasize technology prospecting and market intelligence. This is essentially a search for information, with producers and researchers scanning research elsewhere and looking out for potential market opportunities. Traditional research and experimentation need play little part.

For small countries able to move rapidly, phase II is the normal entry point to the development cycle. They can take advantage of markets that are still growing, get ahead of larger competitors (which tend to be slower to enter the market), and avoid the higher costs and diseconomies of scale from which they tend to suffer during both earlier and later phases. Once socioeconomic and postharvest research has established the potential for a new product, existing technologies developed by other countries can be introduced and adapted.

By phase III, the commodity occupies a significant place in the agricultural economy of a small country. There are many stakeholders who demand support from research to maintain and increase the efficiency of production for continued profitability. However, competition from larger countries is mounting. Research policymakers and managers must decide whether or not to continue their investments in research and, if they do decide to continue, what kind of research is needed. For many small countries, continuing research incurs prohibitive costs, so pulling out seems to be the only realistic option. There is another option, however. This is to continue to do research, but of a different kind. Once an export crop has sunk in value, it can sometimes be transformed into a high-value crop again, through marketing and postharvest research (see Box 6.4).

Linking research investments with phases in the product cycle is important. The crucial decision is when to leave the cycle. The country must pull out before the

---

Box 6.4. Product Transformation

Traditional exports such as cocoa were once exotic, high-value commodities. Early in the 20th century, small countries like Trinidad and Tobago and São Tomé e Príncipe were among the world's major producers. Production shifted from the very small countries to medium-sized ones such as Ecuador, Ghana, and Côte d'Ivoire. Later it moved to the larger countries, such as Brazil and Indonesia, where it is increasingly concentrated today. This shift occurred as large markets and growing world production increased the potential for economies of scale.

Small countries cannot compete with large ones on the basis of price, but they can transform these traditional bulk exports into high-value products once again, this time for specialized markets. The cases of "Blue Mountain" coffee from Jamaica and specialty "fine" cocoas from Madagascar or Trinidad and Tobago show how marketing research and development can create a special niche and command a premium price for a traditional export crop. They illustrate the central role that marketing research plays.

demands on research begin to exceed the limited scale and capacity of the national research system. The warning signal to quit may be one of several. Either the product does not develop further because of market saturation, as in the case of cardamom, or consumers' preferences may change, as occurred in the case of red anthuriums in Mauritius. The most likely signal will be increasing competition from new producers, as occurred in the case of melons. In these situations, research must be able to shift quickly to new priorities.

Figure 6.2 summarizes the role of research in small countries at different phases in the development cycle for nontraditional exports and indicates when key decisions must be made.

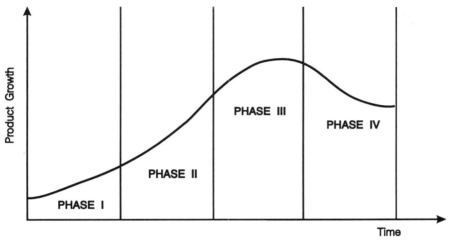

Figure 6.2. Decision-making strategy for nontraditional exports

Managing linkages and scanning information are the key functions in phase I. Phase II involves active prospecting and scanning of available technologies and careful surveys of the market. In phase III, government or parastatal research organizations are expected to provide research inputs along the lines of traditional commodity improvement programs. This type of activity may be inappropriate, and shorter-term partnerships with private-sector producers may be the more effective response. Phase IV is a final chance to transform or abandon the product and avoid higher research costs and declining prospects for product growth.

INSTITUTIONAL RESPONSIBILITIES IN THE PRODUCT CYCLE

Effective research on nontraditional exports depends upon establishing complementarity between private-sector research and development and policy support for research in the public sector. Low-value staple food crops are of little interest to private companies and the impetus for their improvement rests largely with the public sector. In contrast, diversification into high-value commodities is largely led by the private sector, with public-sector agencies called on to provide support. A key role for public-sector institutions is to relieve private producers of some of the high-risk burdens and entry costs that are associated with high-value exports (Crucifix and Packham 1993). Understanding this fundamental difference in the function of public agencies is central to developing an appropriate research strategy for these commodities.

Table 6.2 illustrates how the roles of different types of institutions in the research process for nontraditional exports are distributed over the product development cycle. Because of the changing nature of the demand for research, different institutional actors are more important at different phases of the cycle.

In the early phases, research and development tends to be producer-led. This is a process of trial and error with new crop varieties and management practices. Byrnes (1992) has aptly described this first phase as the "school of hard knocks", in which individual producers take risks, standing to gain a lot or to lose everything by staking their savings on the development of a new product and its initial marketing. When small-scale producers are the key actors, either in a producers' association or as individuals, donor or public support is useful in narrowing the range of technologies and varieties to be tested and in facilitating the technology-transfer process.

Private-sector organizations, including multinational companies, producers' associations, and parastatal commodity organizations, can all play a leading role in the early phases. Farmers and the private sector in the Caribbean, for example, are at the

Table 6.2. Product Phases and Institutional Focus

	Phase of Product Development	Institutions	Research Focus
I.	A nontraditional export is introduced or developed	multinationals, national agribusiness, producers' organizations; policy as facilitator	market intelligence, postharvest and handling, technology prospecting and testing
II.	A viable market niche is found; producers, marketing, packing, and transport systems in place	public and private institutions divide research responsibilities; private-sector research continues to lead, with increased public-sector involvement in research support services	management of pests and diseases, production systems, and farming systems research
III.	The commodity becomes an established export with long-term prospects	a public, private, or parastatal research institution assumes long-term responsibility for the commodity	institutionalized commodity research program: varietal selection and improvement, agronomy, integrated pest management, plant protection
IV.	Competition increases, prices fall, market shares decline	public research policy; producer organizations study markets to decide future of the product	product transformation; or agricultural sector and research policy analysis

cutting edge of technology for nontraditional exports, far more so than government researchers (Walmsley 1990). In Honduras, multinational fruit companies were the first to enter the field and were closely followed by the parastatal IHCAFE. Government research institutions tend to be involved only in the larger, middle-income developing countries such as Colombia, Thailand, and Kenya (Rojas 1993). The voluntary sector also has a role to play. Prospecting for traditional crops that can be transformed into high-value export crops is a task some nongovernmental organizations appear to be doing particularly well.

Public-sector support to research and development at the early stages can take the form of credit or funding to secure technical expertise. In many cases, this has come from donors. Special projects, such as UNCTAD/FPX in Honduras or COLEACP in West Africa, have contributed much to developing high-value exports through (a) trade promotion schemes to introduce new products into northern markets, (b) information exchange between producers and importers on their respective needs and requirements, and (c) technical assistance in the postharvest storage, handling, and marketing operations in developing countries (Guichard 1985).

Sharing of technical facilities, including scientific information services or plots on a research station, may also provide private producers' associations with vital resources to support the early stages of involvement in a new product. Mechanisms and bodies that act as catalysts, linking the public and private sectors and promoting joint ventures between them, can become active. Research foundations, such as the Fundación Hondureña de Investigación Agrícola and the Jamaican Agricultural Development Foundation, prospect for new technologies and transfer them to relevant research groups (Fernandez 1992; Wilson 1992). They often do research on behalf of private producers who are not yet organized as a group to fund and manage research and technology transfer in their own right.

Regulatory and quarantine procedures need to be applied to the technology prospecting and transfer activities of producers and NGOs. As the cycle progresses, the links between public- and private-sector research become better established. Private-sector agencies and producer groups require ongoing policy support and guidance so that the overall impact of new commodities and technologies is not harmful to the environment, or to society as a whole. The functions of quality control and certification become routine and under public-sector management.

In the later phases of the cycle, the commodity has an established group of producers and stakeholders who exert a demand on government research organizations to address the growing range of pest and disease problems affecting the productivity of the new commodity. In addition, increasing competition for markets and the commodity's importance within the country's export economy may lead to demands for agronomic and crop improvement research to reduce the costs of production. Given the volatility and small size of markets for nontraditional exports, public and parastatal research organizations need to consider carefully any investments in longer-term applied or adaptive research. This may be the point at which it is best for a country to move on to another product with higher growth prospects and lower research costs.

Very few small countries are able to undertake the research needed to remain competitive in the final phase of the cycle. Marketing and socioeconomic research are required to identify new opportunities for producers to shift to a new commodity with minimal losses, or to transform the product. Policy is a crucial input into the decision-making process at this stage. Decisions must be made to guide the transition or to determine if the commodity has now become a "traditional" and strategic commodity within the agricultural export economy, in which case a traditional commodity improvement program may be justified.

CONCLUSIONS: CONDITIONS FOR SUCCESS

Research leaders and policymakers in small countries need to be aware of both the potential and the risks of high-value, nontraditional exports. As small open economies, the future of such countries depends on finding new products. Agronomic and biological research is one way of developing such products but is, on the whole, not cost effective for a small country. Other types of research (on market opportunities and postharvest technologies, for example) will allow small countries to borrow technologies wisely and to time their entry into the market in such a way as to lower the risks. Nontraditional commodities require nontraditional research strategies, conducted in partnership with producers and private companies. To take advantage of the opportunities these commodities provide, national research systems in small countries need to consider a broader range of functions, including technology transfer, high-level coordination, regulation, and the provision of information. Marketing and postharvest research are crucial.

The national research portfolio should be able to ensure the following conditions:

- *Swift, flexible organizational responses are essential.* Few small, developing countries can afford the R&D needed to generate new products by experiment, nor can they compete in the market by achieving economies of scale. Their strategy should emphasize rapid entry into the market. By acting swiftly, a small country can enter the market before competitors saturate it and lower the value of the product. It must also be prepared to pull out as swiftly as it entered, once prices start to drop (Hobbs 1988). Rapid, flexible responses are something that small research organizations working with producers can achieve.
- *Producers should play a leading role in research.* Initial experimentation on nontraditional exports should be led by farmers. The involvement of the private sector at the outset is a sign that market potential exists and that private enterprise is strong enough to take advantage of opportunities should they prove attractive. Efforts to diversify that are entirely government driven are rarely successful.

- *Strong links to the policy environment are vital.* Although the private sector should lead the way, the need for advice on which commodities to select from a vast array of potential candidates, combined with the risks involved in technology transfer, mean that strong links between producers and policymakers are essential. The many actors involved in introducing and establishing a new product require good coordination and regulation. Guidance and facilitation will be the major functions of the public sector.
- *Steer clear of product development.* Normally, the investments and research capacity needed to create new products are beyond the abilities of small developing countries. However, these countries should watch for opportunities to transform traditional crops such as *kava, maté, taro,* and *findo* into exports for speciality markets.
- *Strengthen market intelligence and the capacity to screen technologies.* To take advantage of emerging opportunities, small countries need a sophisticated market-intelligence system combined with a well-developed capacity to screen and test new technologies, assessing both their suitability for local production and their acceptability to (foreign) consumers.
- *Invest in research on postharvest and marketing systems.* Postharvest handling, processing, and marketing are the key considerations in providing technological support to nontraditional exports (Kaimowitz 1991; Rojas 1993; Zacarías 1992). In addition to identifying opportunities for new products, market research can produce strategies to add distinctive value to some of the traditional exports being produced by small farmers. Research on postharvest and marketing systems is a crucial area for further investment.

We conclude, however, with a sense of caution. The role of traditional agricultural research policy and organization has not been a significant factor in this story. More work is needed to determine where the opportunities for developing high-value agricultural exports lie for small countries. A key lesson that emerges from the difficult experience with diversification into high-value exports in the South Pacific is the need to identify complementarities in the agricultural research portfolio (Fleming 1995). For example, complementarities between traditional research domains and those that can support high-value, nontraditional crops is essential. Opportunities exist to exploit complementarities among nontraditional crops that use similar technologies and can be integrated into existing farming systems and marketing infrastructures.

Finally, to avoid situations like the unfulfilled promise of cardamom in Central America, small countries can share information and seek complementarities in their own research and development policies with respect to these high-risk commodities. Where the research portfolio can be managed in ways to maximize such complementarities, there is still no guarantee of success and quick riches, but there is likely to be a reduced risk to producers.

7 Biotechnology: A Strategy for Participation

IMPACT OF BIOTECHNOLOGY ON RESEARCH

New developments in basic biological science lead to new technologies for applied agricultural research.[1] Biotechnology is one such scientific development that has had an important impact on the way research is conducted (see Box 7.1). The increasing number of new techniques related to culturing plant cells and tissues, improved diagnostic procedures for crop and animal diseases, and the identification and mapping of useful genes have become invaluable tools for agricultural research programs, including those in the developing countries (Thattapilly et al. 1992; Persley 1991; Getubig, Chopra, and Swaminathan 1991). There is also the possibility of applying these techniques to the conservation of natural resources by developing disease-resistant crop varieties that would diminish pesticide use and by making long-term conservation of plant and animal genetic material more efficient. However, much of this research is carried out in developed countries, often with considerable private-sector involvement; only a small number of developing countries have made substantial investments in research that makes use of biotechnology (Hodgson 1992).

THE CHALLENGE OF BIOTECHNOLOGY FOR SMALL COUNTRIES

Getting involved in biotechnology requires decisions that many policymakers and research leaders in small countries may find difficult to make, as well as an institutionalized scientific capacity that they may find difficult to assemble. Among the issues that research leaders and policymakers face are *intellectual property rights* (IPR)[2] and *biosafety*,[3] both of which affect small countries whether they conduct biotechnology research themselves or merely transfer biotechnologies from another country.

The complexity and costs associated with biotechnology have led some commentators to caution small countries not to invest significantly in biotechnology research at this time (Herdt 1991). Herdt has estimated that even at the lower end of the biotechnology gradient (Figure 7.1), i.e., tissue culture for the disease-free propagation of horticultural crops, the investment in biotechnology requires equipment,

---

Box 7.1. Agricultural Biotechnology and Agricultural Research

Biotechnology refers to techniques that use living organisms, or substances from them, to make or modify a product, to improve plants or animals, or to develop microorganisms for specific uses. Modern agricultural biotechnology uses recombinant DNA (deoxyribonucleic acid), molecular mapping, and tissue culture to obtain a specific product that can be used in crop and animal production, genetics, food and feed processing, management of natural resources, and control of environmental pollution. Research using biotechnology has four important characteristics:

- It is a group of biological techniques, not a research topic per se.
- Developing biotechnology techniques is usually a long-term activity; however, there are also less complex techniques, such as tissue culture, which can be readily used.
- Biotechnology techniques are costly and resource intensive; they require relatively sophisticated equipment and facilities, as well as well-trained staff.
- Biotechnology is not an end in itself; it is most effective where there are clearly defined research objectives and goals in which the role of biotechnology is defined.

Building a national research capacity in biotechnology means

- assembling adequate technical, regulatory, and international expertise;
- setting a balance between technology transfer and building indigenous capability;
- developing regulatory protocols addressing national and international concerns;
- securing resources within a highly competitive funding environment (Cohen 1993).

---

infrastructure, and materials that are more complex and costly than existing research facilities. Despite the apparent constraints that limit their involvement in biotechnology research, small countries will be affected and cannot afford to remain on the sidelines in global biotechnology developments.

GETTING INVOLVED

Some small countries can, and do, invest scarce resources in biotechnology. Honduras and Costa Rica already conduct significant research using biotechnology. Jamaica has invested in improving its capacity to do biotechnology research, while Mauritius has formulated a national strategy for biotechnology. In Burundi, the Faculté de Sciences du Université de Burundi, the national institute ISABU, and the regional institute IRAZ are using biotechnology in breeding, micropropagation, and germplasm conservation. Rwanda's national institute ISAR was using biotechnology in the micropropagation of bananas, sweet-potatoes, and cassava before the civil war. They were even working on meristem cultures and mutagenesis of rice. The Republic of Congo, as with many small countries, is focusing its biotechnology applications on producing disease-free planting material for horticultural crops, tubers, and plantain (Mulongoy 1993).

The main question is not *whether* a small country can develop a research capacity in biotechnology, but, rather, what form it will take. All small countries need the capacity to evaluate the scientific and policy implications of biotechnology that will affect their agricultural and natural resource sectors. For example, in developing

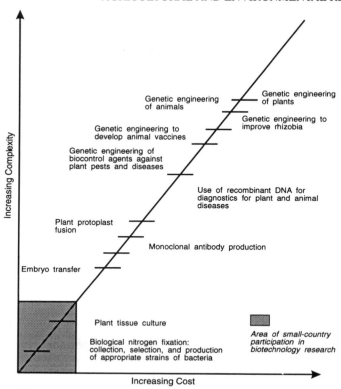

Source: Persley (1990, adapted from Jones [1990]).

Figure 7.1. Gradient of biotechnologies

countries that depend heavily on agricultural exports, biotechnology developments in more-developed countries may have a negative effect by making possible the substitution of their export crops (for instance, sugar) by products from temperate-zone crops (for instance, high-fructose sweeteners from maize). The market implications of these developments need to be analyzed before new investments in a particular export crop are made.

Since biotechnology research often requires additional equipment and skills, countries are faced with difficult choices about how to marshall the necessary resources. The key questions are: What type of involvement is necessary? What is the point of entry? And how can national research systems organize to apply biotechnology to their agriculture and natural resource sectors? In general, NARS in developing countries can be involved in biotechnology at three levels: they can develop new techniques and applications in biotechnology themselves; they can use techniques developed elsewhere; or they can choose to remain informed about scientific and policy-related developments in biotechnology.

IDENTIFYING POINTS OF ENTRY ALONG THE BIOTECHNOLOGY GRADIENT

Biotechnology consists of a gradient of technologies ranging from relatively simple ones, such as plant tissue culture and embryo transfer, to the more sophisticated areas of genetic engineering of plants and animals (Persley 1990). Countries can assess their constraints in the field of agriculture in the light of this gradient, identifying possible biotechnology applications to remove these constraints and hence their points of entry along the biotechnology gradient. The gradient rises steeply, as do the costs and complexity of biotechnology research. Figure 7.1 shows those areas in which small countries are most likely to apply biotechnology to research. However, policymakers and research leaders need to be aware of the entire continuum of biotechnologies that may affect the production and markets of their country's export commodities.

STRATEGIES FOR INTEGRATING BIOTECHNOLOGY INTO THE RESEARCH PORTFOLIO

Deciding whether to incorporate biotechnology into the national agricultural research portfolio—and how to do it—begins with a policy dialogue that links science policy with agricultural development policy. On the one hand, biotechnology applications need to be aimed at resolving constraints to agricultural productivity and natural resource management. This is where agricultural development policy serves to orient the biotechnology applications of highest priority. On the other hand, microbiology, as well as the responsibility for biosafety and issues of intellectual property rights, are likely to be under the purview of science policy. Typically, the bodies that oversee science policy are national councils of science and technology that cover biological, physical, environmental, and socioeconomic research that takes place in universities, research institutes, and ministries (including ministries of agriculture). The first step is to establish a dialogue between science policy and development policy that includes institutions involved in teaching as well as those conducting development research.

A small country's biotechnology strategy must therefore consider the range of institutions that can be involved. Through consultation among scientists, users, and planners, an organizational mechanism can be established to bring the institutions together in a national biotechnology initiative or program. The dialogue normally addresses four questions. One is whether NARS can apply biotechnology in their research programs and how they should go about doing it. The second is how new developments in biotechnology affect the demand for agricultural research services. The third is to identify the policies and regular procedures that are necessary in order to gain access to biotechnology and to apply it. The fourth is to identify the priority areas where biotechnology can provide cost-effective solutions to research problems, along with strategies that pool resources.

Once a decision is made to apply biotechnology in agricultural research, policymakers and research leaders need to identify those domains where the NARS has been able to sustain applied research for commodity improvement. They also need to determine whether there are national scientists trained in microbiology, in either research or educational institutions. In larger or wealthier countries, decisions to apply biotechnology may be made by a single institute, university, or company; in a small country, however, biotechnology applications will normally be the result of national-level decisions involving the entire NARS. Decision makers will need to consider the entire national research portfolio in order to develop a biotechnology strategy.

Strategy formulation begins with a policy dialogue, assesses possible biotechnology applications, sets the scope and priorities for biotechnology programs, and allocates responsibilities among institutions.

INITIATING POLICY DIALOGUE

Even if universities and private companies are the primary users of biotechnology, government remains a crucial partner in biotechnology development. Public policies on education and training, research funding, product regulation, biosafety, patents, and technology transfer determine the degree to which biotechnology can be applied and used effectively (Burrill and Roberts 1992).

Clear advice from research leaders is a major output, from the first policy-formulation phase for incorporating biotechnology in the research portfolio. The goal of the policy-research dialogue is to develop an overall strategy that ensures that biotechnology research is firmly tied to national development needs, rather than to the research needs of special-interest groups or external agencies. A strategy can also help identify the policy issues that need to be addressed. This process of strategy formulation involves matching the country's major constraints in agriculture to the potential applications of biotechnology.

MAURITIUS: AN EXAMPLE FOR OTHER SMALL COUNTRIES

Biotechnology remains a fairly new development in most small-country NARS. As a result, examples of a coherent strategic approach to biotechnology in agricultural research and development are rare. There is, however, one small-country case where a strategy has been formulated. This case illustrates how the national research portfolio was assessed and how a variety of institutional actors were mobilized to apply biotechnology to key problems in agricultural production and resource management.

In early 1992, agricultural research leaders in Mauritius wanted to ensure that their approach to research on biotechnology would be neither fragmented nor ad hoc. In April 1992, with assistance from the World Bank and ISNAR, various institutions participated in a meeting to initiate a more systematic approach to planning and prioritizing their involvement in biotechnology. It was an opportunity for Mauritius to implement one of the main recommendations from the ACIAR/ISNAR/World Bank Biotech Project.[4] That project concluded that "The development of a *national biotechnology strategy*, involving both the public and private sectors is important for implementation of an effective national biotechnology program that makes efficient use of both domestic and external resources (Persley 1990: 66).

The portfolio approach outlined earlier in this book proved a useful tool in formulating this national biotechnology strategy. It enabled decision makers to assess the demand and capacity for biotechnology among a diverse set of institutions. The rest of this chapter focuses on the Mauritius experience, which provides some useful practical guidelines for small countries.[5]

ASSESSING THE CONDITIONS FOR BIOTECHNOLOGY APPLICATIONS

In Mauritius, the necessary conditions to carrying out biotechnology research are present. There is sufficient expertise in conventional agricultural research, such as plant breeding. There is a small but significant community of biological scientists interested in biotechnology. There is demand from industry for technology applications. And, finally, government policy supports the development of science and technology in the country.

Given the diverse institutions interested in biotechnology and the range of objectives and possible applications, a biotechnology strategy was considered to be both necessary and desirable. The first objective was to develop a consensus on the country's strategic goals in agricultural research and the specific role biotechnology would play. More specifically, the following questions were addressed:

- What is the demand for biotechnology applications?
- What type of biotechnology research does Mauritius need?
- Should this research be carried out in-country or can the needed biotechnologies be generated elsewhere and transferred?
- If Mauritius is to engage in biotechnology research, then which institutions will conduct it, how will it be funded, and how will priorities be set?
- What human resource and training requirements need to be met?
- What intellectual property, biosafety, and other policy issues need attention?

A meeting was organized and chaired by the Food and Agriculture Research Council (FARC) in 1992. It included representatives from the University of Mauritius, the

Mauritius Sugar Industry Research Institute (MSIRI), the Ministry of Agriculture, Fisheries and Natural Resources, and private-sector companies, thus bringing together the scientific community, agricultural industry, and policymakers in a national discussion on policy and programs for biotechnology research and development.

The group began by reviewing alternatives and future scenarios for agriculture and natural resources in Mauritius, and reached a consensus on several key points (Box 7.2). Then the discussion focused on identifying the current situation with regard to research in biotechnology and the demands that research in biotechnology may have to deal with.

Box 7.2. Current State of Agricultural Biotechnology Research in Mauritius

Mauritius already applies many simpler and lower-cost biotechnologies in agriculture. The horticultural industry and its supporting research services regularly use tissue culture techniques, while the game ranching and livestock industry use embryo transfer. The sugar industry, historically the largest industry in the country, has supported applied research at MSIRI, whose breeding program has released many improved varieties over the past 40 years and which has recently begun using biotechnology applications.

The University of Mauritius has a School of Science, a School of Agriculture, and a School of Engineering with biological scientists and engineers who can serve as partners with agricultural scientists in a future biotechnology program. The Ministry of Agriculture, Fisheries and Natural Resources also employs biotechnology in plant propagation, and there is considerable scope to increase these applications within horticulture and other ministry programs. Private-sector producers of fish, livestock, and ornamentals are already involved in biotechnology applications that are easily imported from other countries.

The group felt that the sugar industry will continue to dominate Mauritian agriculture; sugarcane production will remain stable in terms of annual tonnage, but if sugar is to remain competitive, production will need to become more efficient. This will require more attention to intensive, high-technology methods and to greater use of the total sugarcane biomass, such as for livestock feed and energy production.

Furthermore, the overall structure of agricultural production will become more diversified and agriculture will contribute less to total GDP, mostly due to growth in other sectors. Agricultural exports will focus on low-volume, higher-value products and on more value-added products from traditional exports. And Mauritius will continue to be a major exporter of ornamentals. In terms of production, less land will be available for agriculture, which, combined with a relative labor scarcity, will raise costs in agriculture.

Natural resources, principally land and water, will have to be used more efficiently as the demand for them for nonagricultural purposes increases. Environmental and conservation issues will also become increasingly important as incomes rise and the demand for improved environmental quality increases.

The biotech strategy group felt that these immediate and short-term developments needed to be considered along with a picture of Mauritian agriculture in 20 years—one that would help them forecast the future problems or issues that biotechnology research will have to address. Even though the state of agriculture and the research problems currently being experienced are widely recognized, it is important to ask if these will be the same in the year 2012. The key constraints identified by the group that are likely to affect agriculture in the year 2012 are shown in Box 7.3.

Box 7.3. Constraints to Agriculture in Mauritius in 2012

By 2012, agriculture in Mauritius will be more diversified and efficient. It will continue to play an important role within a more diversified economy, but key constraints will be
- high costs of production, due primarily to the high cost and scarcity of labor;
- limited land for agriculture;
- the persistence of agriculture on poor soils and rocky land;
- constraints on export development because of seasonal limitations and variable quality of produce;
- small local markets combined with great distances to major export markets;
- choice of crops and stable production limited by cyclones and seasonal availability of water;
- growing pressure on the environment from agroindustry (e.g., pesticides, mill effluents);
- endangering of local species of flora and fauna;
- lack of high-quality genetic material for developing new crops, livestock, and aquatic organisms;
- limited access to information and technology;
- low research investments as percentage of agricultural GDP.

There is much that new technology can do to remove several of these constraints. For example, it can help reduce the need for labor, as well as provide high-quality genetic material for crops, livestock, and fisheries. Through research, the quality of produce can be improved and made more consistent, and its seasonality can be extended. Research can also produce drought-tolerant crop varieties and more efficient systems of irrigation and water management. Biotechnology applications would be useful in many of these areas.

SETTING THE SCOPE AND DETERMINING PRIORITIES FOR BIOTECHNOLOGY

Biotechnology illustrates the dual dimension of scope. Scope is determined by the complexity and intensity of the research (the vertical dimension) and by the breadth of topics and commodities covered (the horizontal dimension). Conducting biotechnology research requires an intensive and complex scientific capacity, which is normally found in countries that have concentrated their efforts on one or two areas over time. Mauritius, which has concentrated on plant breeding in sugar, now has the foundations to apply biotechnology to research. Other countries where traditional plant breeding has been focused on a narrow area are also likely to have a basis for moving into biotechnology research. However, this is likely to be limited to export crops. For many small countries, the possibility of concentrating resources on single-commodity improvement programs has not been possible, and in many cases, it has not been desirable.

A key lesson from the portfolio approach to biotechnology in Mauritius is that once acquired, the capacity for biotechnology research can be applied to a wide variety of topics. The mix and level of disciplinary skills are the crucial aspects; here concentrating or narrowing the scope has been one way to build scientific expertise to a high level. In Mauritius, the use of a portfolio approach allowed policymakers and research leaders to identify biotechnology applications for other domains. At the same time, by understanding the institutional scale and comparative advantage for a biotechnology program, it was possible for them to propose consortia for concentrating and sharing the capacity that was being developed within the sugar and floriculture industries.

The Mauritian group identified two overall goals to which research must contribute and four priority themes for biotechnology research. The first goal is to continue to diversify crop and livestock production with an emphasis on high-value, high-quality products with good standards of uniformity. The second goal is to support the sugar industry by creating more value-added products from sugarcane and reducing the environmental costs of production. The four priority themes for biotechnology research are

- increased availability of disease-free planting material for high-value crops;
- improved quality and control of diseases in livestock and aquaculture;
- increased production of value-added products from sugarcane biomass and reduced environmental impact of the sugar industry;
- continued production of improved sugarcane varieties with desirable qualities such as disease resistance and higher fiber content.

These four priority themes imply the participation of different sets of actors and stakeholders. The key to implementing biotechnology research in Mauritius will therefore be to build flexibility into the organization of the program and to ensure that clients are fully involved from the outset.

ALLOCATING RESPONSIBILITIES FOR BIOTECHNOLOGY RESEARCH

Mauritius is better off than many other small developing countries in that it has a core of highly trained biological scientists in its agricultural research institutions and at the university. Nevertheless, this is a small group that could easily be dispersed or overextended, hence the need to share scientific staff and resources across the institutions contributing to the four priority biotechnology programs.

To this end, four consortia will be organized, each consisting of several institutions but led by the one most concerned and best equipped to undertake biotechnology research for that particular theme (Figure 7.2). Each consortium will have its own operating strategy, enabling it to function as a joint venture among the various institutions involved. It will develop a business plan, specifying the new products and markets at which it is aiming and including budgets, time frames, and indicators of progress (Box 7.4).

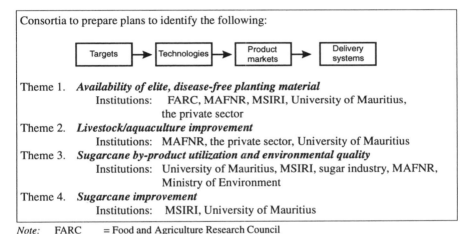

Consortia to prepare plans to identify the following:

Targets → Technologies → Product markets → Delivery systems

Theme 1. *Availability of elite, disease-free planting material*
         Institutions: FARC, MAFNR, MSIRI, University of Mauritius, the private sector
Theme 2. *Livestock/aquaculture improvement*
         Institutions: MAFNR, the private sector, University of Mauritius
Theme 3. *Sugarcane by-product utilization and environmental quality*
         Institutions: University of Mauritius, MSIRI, sugar industry, MAFNR, Ministry of Environment
Theme 4. *Sugarcane improvement*
         Institutions: MSIRI, University of Mauritius

Note: FARC   = Food and Agriculture Research Council
         MAFNR = Ministry of Agriculture, Fisheries and Natural Resources
         MSIRI   = Mauritius Sugar Industry Research Institute

Figure 7.2. Consortia for biotechnology research in Mauritius

At the national level, a flexible strategy needs policy guidance and support to keep it oriented around national priorities and to avoid duplication and the possibility of independent biotechnology initiatives that do not make full use of existing resources. This policy and organization mechanism would be flexible and would act in an advisory capacity. As well as the four biotechnology consortia, the group agreed that a National Agricultural Biotechnology Committee should be established within FARC to advise the Minister of Agriculture on

- opportunities for and availability of potentially useful new technologies;
- policy and institutional issues relating to information access, technology assessment, regulatory requirements, and intellectual property management;
- human resource requirements for biotechnology;
- financial requirements and investment opportunities in biotechnology.

In conclusion, the biotechnology strategy developed by Mauritius is firmly tied to national development needs. It is flexible, cutting across institutions to bring scarce resources to bear on those priority problems where biotechnology can have the greatest impact. It brings together scientists, technology users, industry, and government, focusing their efforts on the areas of productivity and resource conservation considered vital for sustained and equitable national development. For a small country like Mauritius, such a strategy makes the most of the limited resources available for research.

Box 7.4. National Biotechnology Consortia

1. The consortium to increase the availability of disease-free planting material of high-value crops will be led by private-sector companies and associations of flower growers and horticultural producers. Many of the biotechnologies required are commercially available, and the expected returns to the horticultural industry will justify the investment. The transfer of biotechnology will have implications for regulatory and intellectual property issues and will also require access to information and the capacity to assess technology; hence, the participation of MAFNR, MSIRI, and FARC.
2. The consortium on livestock and aquaculture improvement will be led by MAFNR, which is responsible for most of the country's livestock improvement programs. There will be private-sector involvement in biotechnology applications to deer farming and shrimp and shellfish aquaculture. The university's programs will contribute their scientific expertise and laboratory facilities.
3. The third consortium, on sugarcane by-product utilization and environmental quality, will be led by the School of Engineering of the University of Mauritius. MSIRI and the sugar industry will be important contributors. The Ministries of Agriculture and the Environment will be involved to provide policy guidance and support, especially in view of this program's difficult objective of reducing the environmental damage caused by the sugar industry without sacrificing its profitability.
4. The fourth consortium, on sugarcane improvement, will be led by MSIRI, with support from the university's School of Agriculture and the Biology Department.

LESSONS FOR AGRICULTURAL BIOTECHNOLOGY IN SMALL COUNTRIES

Biotechnology is a new item on the agenda of most small-country research organizations, and it differs from the other emerging issues that we have identified here. Instead of being an expansion in scope across research domains and commodities, it is an expansion within domains, mainly through the application of specialized techniques. It therefore poses different problems for research leaders than natural resource management and diversification (Chapters 5 and 6).

Biotechnology policy and planning issues cut across programs and institutions involved in research. Biotechnology applications can place relatively high demands on the scale of research: it is expensive and it needs specialized facilities and equipment and highly trained staff. Finally, biotechnology research involves many diverse institutions, and the only way that small countries can achieve sufficient critical mass is to include all of them, a process that must begin with strategic planning for the priority areas where biotechnology applications are relevant.

Almost all of the countries in Latin America have created national biotechnology programs and defined policies for biotechnology. In many cases, these countries have set up specialized biotechnology research and development institutes; in some (Argentina, Brazil, and Chile), biotechnology programs have been introduced into existing institutions (Commandeur 1993). Establishing independent biotechnology institutes and facilities in small developing countries is rarely feasible; integrating biotechnology research into the breeding programs of established commodity institutes is more realistic. Where biotechnology is not the extension of a traditionally strong breeding program, it may be more appropriate to develop new, flexible institutional arrangements that share resources across institutions and which can focus biotechnology research on a few priority problems. Establishing a national program that shares the physical and scientific resources of several institutions is one way to build the capacity for biotechnology research.

One option for countries facing a demand for biotechnology research that is clearly beyond its scale, is to contract the research out to university or commercial biotechnology labs in developed countries. To some extent biotechnology has altered the need to locate the research in the areas where a plant or animal is to be produced. In the private sector especially, research and development efforts that cross national boundaries are now common. Companies have moved quickly to identify and access technical expertise scattered throughout the world. So far, these exchanges have been between developed countries; developing countries must give serious thought to how they can also participate (Wagner 1992).

Tissue culture is one biotechnology where technology transfer is well established. If a small country wishes to do its own work, a good entry point is the establishment of tissue-culture laboratories. In many larger countries, such laboratories might proliferate within various commercial sectors, but in small countries there may not be any justification for maintaining more than one such facility, which can serve the needs of several commercially valuable commodities and industries. Novel management mechanisms are being tried in Jamaica and Mauritius, for example, to make such facilities available for a variety of commercial and even noncommercial uses.

Since biotechnology is neither a research topic nor an end in itself, considerations of scope have more to do with the build-up of a scientific capacity in agricultural biology than in commodities or themes. Elevating and focusing the system's scope on the level of molecular biology will not be easy in countries where the main pressures on the national system are to provide broad coverage with less experimentation and applied research. For countries such as Mauritius where applied research in one domain has been justified and is feasible, the conditions to apply biotechnology to research are more positive. This does not mean that biotechnology applications need to remain confined to the research program or domain where the scientific capacity was first established. Biotechnology applications can be wide ranging, depending on their complexity and existing priorities. Focusing on priority themes and creating wider access to biotechnologies is essential in justifying the investment.

The assessment of the constraints and priorities for biotechnology in Mauritius using the portfolio approach was useful in identifying the demand and the actors. Several lessons emerge from the Mauritian experience in formulating a biotechnology strategy:

- While the development of new applications and methods in biotechnology is beyond the capacity of many small countries, biotechnology itself is not. Each small country can participate in biotechnology; the decision is whether it will participate as a developer of new techniques, a consumer of existing techniques, or a reviewer of biotechnology applications, and whether this level of participation will be the same in all areas where the techniques can be used.
- Small countries would be wise to select a few priority areas in which biotechnology can make significant contributions. Nothing is to be gained, in biotechnology as in other areas of research, by spreading resources too thinly over a wide range of projects, commodities, and objectives.
- No single institution, private or public, is likely to have sufficient scientific and financial resources to establish a national biotechnology program that addresses all the priority needs. However, various institutions can be brought together to share resources and facilities and to provide adequate coverage of all the issues raised by biotechnology research. The private sector, in particular, can play an important role in biotechnology research and the distribution of its products.
- Policy issues need to be addressed at all points along the biotechnology continuum, from the transfer of existing biotechnologies (such as plant tissue culture), to the development of new ones (including more complex genetic engineering).
- Some biotechnology applications focusing on environmental and conservation issues or on staple food crops have no immediate or foreseeable commercial value. For these, public-sector financing is essential. However, integrating biotechnology into existing research programs appears feasible in only a few small countries at present. These are the ones with a major agricultural export crop that has been well supported by research over several decades. This research is often based within parastatal commodity institutes or research foundations.
- In some small countries, traditional commodity improvement programs are difficult to sustain because of problems assembling a group of scientists focused on one commodity. These countries are likely to have problems applying biotechnology to research. For these countries, policy guidance and regulation of biotechnology transfer will be the most important contributions that research can make.

In Mauritius as elsewhere, there is a strong demand for biotechnology from national programs and commercial farms interested in developing high-value export crops. The applications are for readily available biotechnologies such as tissue culture for the production of disease-free planting materials. A tissue-culture facility that is available to a wide range of users in the agricultural sector is one option that most small countries should consider. There are few issues of property rights or biosafety to complicate involvement.

Private companies are often leaders in seeking out biotechnologies. For the transfer of more complex biotechnologies, including transgenic materials or diagnostic techniques, policy support is needed to deal with property rights and biosafety issues. Countries with universities have an advantage in monitoring biotechnology developments elsewhere, as well as being able to mobilize scientists for domestic research on biotechnology.

Generally speaking, the conditions for success in a small country are (a) a university with a biology department and agricultural sciences, (b) an established breeding program in a major commodity, (c) a private sector that is willing to invest in biotechnology, and (d) an information service capable of linking to sources on recent developments. As biotechnology developments lead to new and more affordable applications, small-country NARS will need to consider formulating national strategies to take advantage of new opportunities that biotechnology can provide.

Notes

1. I would like to thank Joel Cohen and John Komen of ISNAR for their comments and suggestions on this chapter. Their Intermediary Biotechnology Service project has produced useful guidelines on the policy, organization, and management of biotechnology in agricultural research. This information is available from ISNAR.
2. IPR is that area of law concerning patents, copyrights, trademarks, trade secrets, and plant variety protection. Many of the results of modern biotechnology research are proprietary. Private institutions increasingly seek IPR protection for biotechnology inventions, and a similar trend exists for advanced public-sector research institutions in developed countries. Developing-country NARS, which generally rely on public-domain techniques, are increasingly affected, both in obtaining new biotechnologies from outside and in safeguarding rights and benefits to local biotechnologies.
3. *Biosafety* refers to the policies and procedures adopted to ensure the environmentally safe application of biotechnology. The production and release of genetically engineered organisms has raised concerns about potential risks to public health and the environment. Assuring compliance with biosafety regulations is important to foster public acceptance and further development of modern biotechnology. For developing countries, biosafety regulations are also necessary to facilitate access to biotechnology generated abroad. In any country with a biotechnology program, a national biosafety system is essential to regulate production and release of genetically modified organisms (see Persley, Giddings, and Juma 1993).
4. From 1988 to 1991, the Australian Centre for International Agricultural Research (ACIAR), the Australian International Development Assistance Bureau (AIDAB), ISNAR, and the World Bank sponsored a project to assess the potential of biotechnology for contributing to increased agricultural productivity in developing countries. They identified socioeconomic, policy, and management issues that could influence the proper introduction of biotechnology into these countries. The project findings are documented in two books (Persley 1990, 1991).
5. The full report of the meeting is contained in a paper by Antoine and Persley (1992).

8 Planning for the Information Revolution

TRENDS IN THE INFORMATION ENVIRONMENT

In recent years we have seen tremendous developments in information technologies. While each generation of computers decreases in size, their capacity to store and manipulate data increases exponentially. Telecommunications and satellites allow us to communicate quickly with scientists and colleagues throughout the world, and equally important, the costs of the technologies are declining each month. These technological developments are the driving force behind the current enthusiasm expressed by scientists and information specialists alike for applying information technologies in agricultural research. In contrast to the situation five years ago, a research institute that does not rely on computers is becoming the exception rather than the rule.

There is now much more interest in information and its management than ever before among both managers and scientists. Often, this is because of the relatively sophisticated information-handling opportunities available to individual managers and researchers, who are beginning to experiment with the various possibilities they now have access to. And while managers and scientists are learning what they can do with the new computer technologies, information specialists are taking the opportunity to extend and improve their services through computerization.

The increasing technological capacity of national agricultural research systems allows them to be much more sophisticated in the ways that they generate and manipulate information. But these new, sophisticated technologies are not without their negative side. While they provide NARS with the potential for greater access to information than ever before, as well as the tools to manage it, they also create problems: they can overload already overburdened systems by providing "access to more, instead of offering selective mechanisms for limiting supply" (Owen 1989). Part of the problem is that as information technologies spread, the number of institutions that can contribute to the stock of information increases, and the number of information sources to which the NARS has access also becomes greater. Instead of working with a few well-established organizations, a NARS now receives information from more and more sources, both in the traditional research community as well as from a diverse range of other institutions.

CHALLENGES TO SMALL COUNTRIES

This general trend of increasing quantities of information, more diverse sources, and easier access through information technologies is affecting agricultural research institutes the world over. So what does it mean for small countries?

- The smallness of a country can be a major advantage with regard to information technologies. If we ignore the organizational complexities associated with the content of what is to be managed and consider only the technical feasibility of computers and telecommunications, then it is an easier task to computerize the entire research system of Benin or Swaziland than to do the same in a large country such as Nigeria. For a small investment, an entire research institute can have a local computer network, be connected to other institutes in the country, and also have electronic mail connections with the rest of the world.
- In the present economic environment, not much can be done to increase the limited resources of research institutes in small countries. Their only option is to develop innovative strategies that make better use of existing resources. One strategy is to pay more attention to the potential contributions that enhanced information management can make to research.

The main challenge for small-country research and information managers is to identify the information they need from the mass that is available to them and to find ways of getting it. It is not just quantity that is a problem, however: the patterns of knowledge generation and the flow of information are also changing (Biggs 1990). This means that information units have to be reoriented in order to get access to what they need. Similarly, the information demands posed by research in areas such as biotechnology and natural resource management create real problems of scale and scope. How can a tight information budget be stretched to effectively monitor developments in subjects that formerly did not exist, or were considered to be of marginal interest? Small countries lack the resources to maintain in-depth contact with all potential sources of information. Instead, they need to develop strategies that will give them access to the information they need, and there are many different ways this can be done.[1]

The current response of many NARS (and the donors who support them) is to focus on technologies: the equipment, software, and hardware needed to handle information. The aim is to increase efficiency in information access, storage, and dissemination. However, quantitative improvements must be balanced by other measures to ensure that issues of quality, relevance, and local institutional capacity are also considered. It is essential that research institutions look at ways to reorient their information efforts to deal with floods of information rather than to cope with droughts. In small countries, where diverse information demands must be met from limited resources, information managers need to concentrate on information quality and selectivity, rather than increasing the quantity of available information, as was done in the past.

New information technologies can be used to open up new possibilities for communication and access to new sources of information. Installing technologies or getting access to telecommunication links is relatively straightforward, but to make the best use of them, investments in technology must be complemented by investments to establish a basis for screening, evaluating, and synthesizing information so that it is relevant to local need and demands. Developing this capacity requires knowledge of the research system and its activities, familiarity with different information sources and their characteristics, and the availability of a range of information and communication mechanisms to suit the system's situation. Moreover, to be effective, the required expertise and technologies, which are normally scattered across several institutes, need to be linked and coordinated so that individual researchers or institutes can rely on an information system with a capacity to manage and deliver useful information that is greater than the sum of the parts.

Putting this kind of system together requires, first, a good appreciation among policymakers, managers, and scientists of the overall role, functions, and contributions that information plays in agricultural research. Developing an information strategy is also a useful way to involve different parts of the research information system in joint approaches that build on the strengths, weaknesses, and comparative advantages of each institution.

AN INFORMATION MANAGEMENT STRATEGY

Scientific information must be preserved, stored, and made available to researchers, so there is a *storage* or *repository function* to information management. Scientists must be kept abreast of new developments, so there is a *current-awareness function*. And there must be a *scanning function* to identify and track potential sources of information to ensure that new sources are accessible. Determining which function to perform and who should perform it is the purpose of an information strategy (Ballantyne 1993). These are the key factors to consider:

- *Demand* for information. This represents the *needs* of scientists, research managers, and others for specific pieces of information. The nature of the demand is determined by factors within the research system itself as well as in its external environment.
- Information *sources* to satisfy demand. These are organizations and individuals that generate or distribute information.
- Information *mechanisms* used by scientists and organizations to contact sources and obtain information to meet identified needs.

A successful strategy depends on research managers and information specialists working closely together to develop a broad vision of what information is, how it can contribute to agricultural research, and how it can be managed as part of the organization's overall strategy for research planning and management. Drawing from the factors given above, the three elements that an information strategy must contain are identification of research demands, assessment of information sources, and selection of information mechanisms. In addition, whether the strategy is for an institute or a research system, the whole activity needs to be coordinated and managed.

IDENTIFYING DEMAND

The demand for information corresponds to the agricultural production systems of a country, the scope of the research program, and the kinds of research conducted (see Table 8.1). These, in turn, determine the types of information that should be provided by an information unit and how it should be provided. The demand for information varies over time and shifts with each change in research priority or project emphasis. The situation in Swaziland illustrates how the structure of agricultural production influences the research system and the demand for scientific information (see Box 8.1).

A close assessment of the information needs in the NARS should be made to determine the characteristics of the demand, and based upon that, an information system may have to reconsider its commitments and modify its strategy. The growing role of biotechnology in agricultural research is an example of the kind of thing that can cause such change. Keeping up with this rapidly changing field is difficult even in well-endowed institutions (Eaton 1991), and traditional information services such as libraries may not be able to provide the necessary access to up-to-date developments in biotechnology. In fact, scientists in biotechnology often rely on personal contacts and specialized networks for this information. Linking the demands or needs with the ability of information mechanisms to deliver specific types of information can reveal a situation where the focus of a mechanism like a library should be changed to better complement other mechanisms.

Unfortunately, the less-traditional information mechanisms have received little attention from information managers. Documentalists rarely regard research networks or workshops as mechanisms of information exchange, and scientists tend not to think of their study tours as part of an organizational strategy to obtain information. As a result, much useful information is lost or not available to a wider group of users, and resources are wasted in efforts to obtain information that is already available, but through some other source.

One way to get information staff in touch with alternative sources of information is to broaden their mandate and activities beyond traditional documentation. Information specialists must interact closely with scientists in order to be able to design suitable information strategies. In addition, they must monitor the national development policies that determine research agendas and anticipate which programs will need support. If information specialists can develop these contacts, they can become more aware of what is going on and of what is planned for research, allowing their

Table 8.1. Factors That Influence Demand for Information

Characteristic	Influence
Scope of Research	The scope of research determines the kind of information needed by researchers. Because small countries cannot generate all the information that they require from research, they need to acquire it by using information mechanisms. Thus, the *scope of information entering the system may be broader than the scope of the research itself.*
Change and Diversification	The effects of changing policies are more pronounced for NARS in small countries, particularly in response to external economic factors. Changes in research focus also require that their information support be reoriented to serve new needs. This involves *links with new sources of information and possibly new mechanisms.*
Production Systems	The orientation of research depends on the farming systems where new technologies are to be used. Differences between farming systems create demands for different information, which can be met by *tapping sources serving similar production systems.*
Kinds of Research	Small countries are more involved in research on adaptation and testing, requiring a capacity to scan upstream knowledge looking for suitable technologies to acquire and adapt. *Information mechanisms need to be broadly based and able to access a wide spectrum of information.*
Socioeconomics and Farming Systems Research	Socioeconomics and farming systems programs need access to information that is generated and collected locally, since regional and other external agencies have little ability to do so. These kinds of information are often generated by scientists and institutes outside agriculture: *useful local information is available outside agricultural research institutions and different patterns of information flow can be expected.*
Time Dimension	Research on perennial crops or natural resources requires long-term information commitments as compared to that on annual crops or other short-term research priorities. *Frequent changes in small-country research programs may cause problems for long-term projects; information support over a long period requires special treatment.*
Research Personnel and Their Functions	In small countries, limited numbers of research staff can make it easier for information services to identify and monitor their needs. However, movement of staff between crops or commodities alters their needs for information, and frequent monitoring of staff activities is needed. *Researchers in small countries are frequently involved in other tasks, and their demand for information may reflect these needs rather than their research activities.*
Institutional Structures	*The nature of small-country research systems, with few specialists and many generalists who work on several crops or commodities, makes it more difficult to target information services.*

Source: Ballantyne (1991).

services to be adjusted to changing demands in a timely way (Hee Houng and Ballantyne 1991). With no knowledge of the research programs and of the pressures that shape research agendas, an information system can quickly become irrelevant.

> **Box 8.1. Swaziland: Farming Systems and the Demand for Research Information**
>
> Research on Swazi nation land, where small-scale subsistence agriculture is dominant, is the responsibility of the government's Agricultural Research Division (ARD). By way of contrast, on individual title deed land, which is dominated by cash-crop production, commercial companies conduct their own research or depend on technologies from outside sources. The result is two parallel research systems focusing on different crops and working with different types of producers. Their information needs are, therefore, also different, as are the sources they rely on and the information mechanisms they use. In the ARD, research networks affiliated to international research centers provide most of the information and technologies needed for research on staple crops such as maize and beans. The commercial companies producing sugar, however, rely on their personal contacts with experts outside the country (mostly in South Africa) and their memberships in professional or technical associations. In this situation, where researchers have access to much of what they need, the potential role of traditional libraries needs to be clarified, perhaps through an information strategy.
>
> *Source:* Mavuso and Ballantyne (1992).

ASSESSING INFORMATION SOURCES

There are many potential sources of information, and as more organizations become involved in research, relevant information becomes more diffuse and scattered and more difficult to identify and obtain (Biggs 1990). Table 8.2 summarizes the characteristics of different types of information sources.

No single organization, especially in a small country, can house and access all the information that might be useful to it. Instead, it must pursue a strategy that gives it a broad awareness of potential sources, which should be combined with a specific focus and careful selectivity in acquiring information. There are international organizations such as CABI, FAO, and CTA that provide secondary access to information through digests, reviews, abstracts, question-and-answer services, data bases, current-awareness services, syntheses, and referrals. They provide prescreening and digesting services that can be particularly useful to small countries, but at a price, which is important in determining the degree to which small, poor countries can use them.

Small-country research systems must pay close attention to the different patterns of availability and information flow between commodities and the institutions that work on them (Box 8.2). These patterns determine the amount, quality, and usefulness of available knowledge. Identifying, characterizing, and assessing different sources requires that the information specialist develop particular skills: a broad awareness of sources of agricultural information and an understanding of institutes is required, as well as current information on their programs. Some of this can be obtained from documents and reports, but other forms of in-depth contacts are needed to make the best use of key sources.

Assessing the usefulness of external sources of knowledge to in-country programs requires accurate information about local requirements and demand. This highlights

Table 8.2. Sources of Information: Characteristics and Issues

Information Source	Characteristics	Information Issues
International Centers	• publicly funded • oriented towards developing-country problems • often coordinate networks • research moving upstream	• relatively easy access • specialized subject focus • large output of knowledge • use variety of mechanisms to disseminate information
Regional Centers	• publicly funded • most are oriented towards small countries • often coordinate research rather than doing it themselves • research tends to be adaptive and applied	• relatively easy access for countries in the region • often a focus for information development and coordination for the region • more relevant outputs
More-Developed National Systems	• publicly funded • moving to commercialization • research tends to be applied and strategic	• relatively easy to identify information • costly to obtain information • information must be adapted to small-country situation
Less-Developed National Systems	• publicly funded • donor supported • research tends to be location-specific	• difficult to identify and retrieve information • limited knowledge outputs • outputs often very relevant
Universities	• publicly funded • research tends to be upstream • more socioeconomic research • research is only one role of staff	• good sources of in-depth, wide-ranging information to support longer-term research • different method of access to upstream research • good source of interdisciplinary information • major source of socioeconomic data and knowledge
Private-Sector Organizations	• research tends to be applied • research oriented to commercial applications	• dissemination of information restricted • information usually limited to situation or product • information costly to obtain • information difficult to identify

Source: Ballantyne (1991).

the need for close cooperation between the people who scan and assess potential knowledge and those who will use it. Information management in small countries needs to be organized so that the demand profile used to evaluate external sources of knowledge is kept up to date with the needs of researchers.

While information must be relevant, there are other criteria for assessing sources of knowledge. These include the direct and indirect costs of acquiring the information,

---

Box 8.2. Trinidad and Tobago: Coping with Change

Trinidad and Tobago is a case where the information service is prominent and well connected. Providing information for more than just research, it is based in the ministry of agriculture where it serves policymakers as well as producers. With such a mandate, information managers are well placed to interact with research leaders and bring the information function into the core of research management and planning.

Times, however, are changing rapidly in Trinidad and Tobago. The country is diversifying its agriculture away from the traditional focus on sugar, tree crops, and food staples. And as this happens, the ability of the central information service is being stretched to access, store, and disseminate the wide range of information that is needed. As researchers cope with the changing scope of research, they are looking for other sources of information, employing new, less-formal mechanisms and gaining access to rapidly changing information sources. The current centralized information mechanism cannot hope to cover this new breadth or deal with so much timely but ephemeral information. It should, however, be able to assist in identifying and cataloguing information sources so that the information that enters via visitors and networks can be made available to other users. A strategy can also be developed to avoid duplication of the central storage and dissemination functions of the Ministry of Agriculture's libraries by other research units in the country.

Source: Hee Houng and Ballantyne (1991).

---

the speed with which it can be delivered, its user-friendliness (in language and format), the resources needed to adapt it to local needs, and last but not least, the credibility of the information source and its products.

SELECTING INFORMATION MECHANISMS

Once information needs have been identified and potential sources of information have been assessed, the most suitable mechanisms for obtaining the information can be selected. Depending on the situation, these mechanisms could include libraries, information networks, research networks, meetings, study tours, visitors, or personal contacts. At present, little conscious selection of information mechanisms goes on in agricultural research institutes; most organizations tend to continue with traditional means. The result is that some mechanisms are expected to provide information that they are unable to deliver, while others (that could deliver what is needed) remain unused.

Research organizations or systems often expect a single information unit to carry out all essential tasks, such as identifying information needs, assessing and evaluating available information resources in relation to those needs, influencing external agencies that produce or deliver information, and delivering the services and products that scientists and managers need. In most situations, the requirement is so large and involves such a variety of mechanisms and sources that no single information unit could possibly carry out all the tasks effectively. A combination of mechanisms is needed, each with a different mandate, depending on its comparative advantage in either serving specific research needs or accessing information sources.

Libraries or documentation centers exist in virtually every agricultural research center or institute. They range in size from a single bookcase to large, well-organized collections of reports and journals backed up by computerized databases. Agricultural research libraries act as an information service for research and, in small countries, often also act as the national depository or archive for agriculture in general.

The libraries of small countries have multiple roles, functions, and clientele. Most struggle to do a good job of storage and have difficulties with dissemination. These two functions are quite different and attempting to do both may stretch a small information service in potentially conflicting directions. These libraries attempt to serve diverse groups, including policymakers, researchers, students, extension workers, and even farmers. Large countries can afford information services for each group of clients, and they usually have sufficient staff to support specialized centers, each with a different role. If these multiple roles and client groups are to be served through a single unit in a small country, however, it is important to recognize the pressures on a small facility attempting too many tasks with inadequate resources.

One concern that was consistently raised in the ISNAR case studies was the loss of information produced within the country. Small countries, of necessity, depend on the work of donor projects, which produce much useful information but tend to invest nothing in making that information available to a wider community of scientists. In many cases, the information service needs to carry its storage and retrieval function one step further, perhaps to produce the documents themselves in a small publication unit.

Information networks exist at the national, regional, and international levels. The information units involved in these networks collaborate mostly in training and interlibrary exchange, but also in developing tools and standards for data sharing and access (Foss 1990; Sison 1990). Most information networks are extensions of formal library and documentation services that have enlarged and enhanced the services available to researchers in participating countries. They rarely change in any major way the kinds of information that a researcher needs, nor do they really open up new information sources. Their main contribution is to provide more efficient *access* to the kinds of information that already exist in documentation centers.

The number of networks an organization participates in, and the degree of activity in each, needs careful management (see Chapter 10). If network activities can be shown to enhance local services, which is the case with many of them, then the efforts to participate can be justified. Care must be taken, however, to ensure that national efforts are not consumed by activities designed to strengthen some external entity.

One of the advantages of network participation is that it can save time and resources. But this can only happen if there is enough scope for a division of tasks and labor. Where there is too much similarity between participants' activities and resources, as was the case of Cape Verde (Morais 1992), the objectives of the network are often defeated.

The greatest strength of *research networks* as mechanisms for delivering information is that they can provide current and highly credible information. Network coordinators are usually familiar with the researcher's program and understand the problems being studied. This helps them target information to individual researchers, and the information in turn is seen to be very useful and credible because of the coordinator's expertise. Coordinators are usually based at an international or regional research center, and they are in a good position to monitor current developments, disseminating important information to the network. Although much of this information is available from other sources, it is often perceived to be more useful and dependable if it comes from the network.

Other positive features of research networks are the range of communication methods available. Knowledge is spread through newsletters and publications, but regular visits by the coordinator, along with training courses, conferences, and study tours arranged by the network, are probably more effective. These opportunities are not regularly available to a research system that relies on libraries alone; meetings with other scientists are often unplanned and infrequent and cannot be relied upon to deliver the knowledge that is required.

Personal contacts and meetings are a major way that information enters research organizations. In the South Pacific, study tours, visitors, and attendance at meetings are identified as a key means for researchers to acquire information (Sivan 1992b). Donors and technical-assistance agencies support many such opportunities as well as more formalized exchanges via agricultural research networks (Burley 1987; Plucknett, Smith and Ozgediz 1990). These networks are channels for distributing up-to-date information to participating scientists. Professional associations and societies represent an older form of networking for individual scientists, offering much the same kind of benefits to members. Scientists are also well known for their personal networks, the so-called "invisible college".

Scientists will always rely on personal contacts for large amounts of their information needs. Managers can promote these contacts by providing and supporting opportunities for staff to attend meetings, study abroad, or go on study tours. Similarly, they can be open to outside expertise, encouraging movement and contacts between staff of different institutes who work on similar problems. The disadvantage of personal contacts is that the information and knowledge acquired may never be shared with other staff in the system, nor does it become part of the institutional memory, available for future decisions and research.

As small countries begin to adopt more flexible approaches to agricultural research and development and to respond to new agendas such as environmental conservation or biotechnology, no single mechanism for acquiring information will be able to satisfy all their needs. Even a micro-state such as the Seychelles will need to keep several channels open for information on research and development. In the Seychelles, small, decentralized documentation centers have been set up to ensure that agricultural information is stored, retrieved, and disseminated. These centers are intended to provide an institutional memory (especially for locally produced information) that did not exist in the past when scientists relied almost exclusively on personal contacts and visiting consultants. However, personal contacts between scientists remain vitally important in the Seychelles as a source of current, specialized information that cannot be easily obtained through formal channels (Mend and Ballantyne 1992).

Selecting the blend of information mechanisms is not easy for small countries. Information has a broader and more important role to play in a small country than it has in a larger one, so all kinds of information are needed and all types of information sources must be accessed. Their scale, roles, and functions, however, must be different than those in larger countries; small countries need to experiment with different combinations and approaches, and even new models may be appropriate.

The degree to which small countries can afford to maintain active participation in research networks as well as investing in more formal libraries or information units is questionable. The decision is not *either* networks *or* libraries, but what investment should be made in each and for what purposes. The particular balance will vary according to local needs and the characteristics of the flow of information for the subject being researched.

When information mechanisms are selected, the demands to be met, the sources to be tapped, and the options presented by other mechanisms should all be taken into account (Table 8.3). This requires a systems view, where each component is not only aware of its own functions and resources, but where it is also aware of its functions *in relation to* the functions and activities of the other components. Such a holistic view considers the contributions and roles of all the institutions that make up the research system, as well as the roles of the different information mechanisms. It also ensures that the responsibilities for an information function will be assigned to the institution or mechanism that is best equipped to carry it out.

It is essential that decisions on what mechanisms to use are broadly based and that they take into account what each mechanism can contribute to a total information strategy. For example, services from a library are more effective if they complement the activities of networks or professional organizations, rather than trying to provide the same information. A comprehensive information strategy would allow screening of the various sources of information, including networks, visits, and meetings. It would enable the organization to exploit the main advantage of networks for example (that they are timely and well targeted), while at the same time avoiding the disadvantage (that much useful information stays with network researchers and does not become available to the institution). It would also allow the organization to screen for those bits of information that are relevant to its overall mission and then to circulate and store them.

There are obvious parallels in a strategy for information and a strategy for the research system as a whole. In research we have noted that no single organization can cover the full scope of the demand for research information and technology for the

Table 8.3. Information Mechanisms: Characteristics and Uses

Mechanism	Positive Features and Uses	Problem Areas
Scientific Information and Documentation Services	• access to traditional documentary sources • broad subject coverage • institutional memory/repository function • access to public sector, universities, IARCs, and regional centers	• often out of date • not so good for nonbibliographic sources • poor links with private sector • emphasizes quantity more than quality and synthesis • more process oriented than use oriented • slow response to program changes
Networks	• access to traditional documentary sources • access beyond local resources • access to public sector, IARCs, and regional centers • fosters collaboration and extroverted attitudes	• requirements to participate may overload small systems • stability and sustainability problems with donor funding • multiplicity of networks
Personal Contacts	• very high credibility • specialized subject coverage • all types of sources • convenient • multiple functions and purposes	• personal bias • problems with redisseminating and using information in the system • difficult to manage

Source: Ballantyne (1991).

agricultural sector; likewise, no single information mechanism can satisfy the range of information needs within the research system. The key to success is to have several mechanisms to tap into a range of sources. In practice this is often what happens; however, the case studies showed that this diversity is not always managed to best effect. Where there are multiple documentation units (in institutes, councils, ministries, and universities), there is a tendency for each to compete with the others, going at times to the extreme of photocopying each other's collections to demonstrate their broader coverage, when, in fact, there may be different users and uses of the documentation in the different centers. Paying more attention to identifying the users and their needs would soon make it apparent that there is a greater basis for complementarity than for competition and duplication.

IMPLEMENTING A STRATEGY

While it is relatively easy to identify the main elements of an information strategy (Box 8.3), it is more difficult to identify who should be responsible for its development. The problem in most small countries is that each institution has its own information personnel, each with different institutional mandates. Very rarely is there an institution

---

Box 8.3. Features of an Information Strategy

- Recognition by managers and policymakers that information is a valuable resource
- Characterization and assessment of research demands
- Assessment of potential sources of information
- Identification of different mechanisms for acquiring information
- Coordinated division of labor among several participating units or institutes
- Clear and strong linkages with research activities and structures
- Differentiated use of information mechanisms according to demands and sources
- Involvement of different information and communication personnel

---

with a national mandate in research information. Where there are national agricultural information or documentation centers, their mandates usually cover all of agriculture and are not limited to research, which means that the all-important coordination of a multi-institutional approach is very difficult to achieve. While individual institutions are keen to work towards a national plan, they have no authority to call meetings or organize specific activities except on a voluntary, nonbinding and informal basis.

Since agricultural research itself faces the same problem in most small countries, it is heartening to see the emergence of research coordinating bodies in several countries. A link between research coordination and planning and information planning may be an effective way to overcome the problem of responsibility in the information sector. In Sierra Leone, the agricultural research strategy developed by the National Agricultural Research Coordinating Council (NARCC) already has an information component, which could be a model for other countries to follow. It would have the benefit of linking information planning with research planning and offers opportunities for the constructive dialogue between research managers and information professionals that is so important.

It is very important in a small country for several institutions to be involved in information-planning exercises. This allows participating institutes to take on specific responsibilities for the group, naturally allowing others to reduce their efforts in the same areas. The aim is to achieve a division of labor in which all participants have clear tasks, roles, and contributions, rather than trying to centralize all the tasks in one center, which would not have the capacity to do them all.

A holistic approach to the information system implies both variety in institutional representation as well as variety in types of information mechanisms. Coordination of libraries and documentation centers is in itself valuable; it is even more effective if it complements the activities of networks or other kinds of communication mechanisms. Broadening the dialogue to involve network coordinators or others involved in data-base creation, telecommunications, or editing and publishing provides a better basis for an information strategy that can respond to the widening demands of researchers in small countries.

CONCLUSION: OUTSTANDING ISSUES

In the past, information management and documentation were considered to be support functions, secondary to the core function of agricultural research: experimentation. In this chapter, however, we have argued that information management is a central function of a small-country NARS. Now, *managing* scientific information is acknowledged as being essential. By gaining access to information and research results from other countries or from global sources, NARS become better informed and more capable of performing their advisory and regulatory functions; by scanning external sources of information, they can focus their research on problems for which technology and information are not available.

This shift of the information function from a supporting role to the core of agricultural research has managerial implications for NARS that stretch beyond the details of technologies and service delivery. Because the availability of scientific information has become a major factor in determining the feasibility of a research program, information management must be included in the planning and design of research programs. And information managers need to work together with program managers to implement this approach.

In many small countries, information management is not yet recognized as a key activity and function of research institutions. In some cases, parts of the information function have been identified and they are visible as information units within the system. However, other activities, such as attending meetings or participating in networks, are not considered to be informational, even though their main function is to acquire knowledge and information, and they are not managed in any systematic way. It is essential that the information function be recognized and identified before an approach to acquiring and managing information is developed. An incomplete view of the information function and its parts will result in inadequate access to information.

CHANGES IN CONVENTIONAL ROLES

As research in small countries becomes less experiment oriented and involves more knowledge assessment and synthesis, the role of researchers will change. Differences between researchers and information staff should become less distinct: knowledge handling will be recognized as the main task of both groups. Each group would, of course, continue to bring different skills to a problem, but researchers would be expected to play a clearer and more active role in assessing and repackaging knowledge. In small countries, researchers will require some specialized training in information or knowledge handling to prepare them for this role, a role that will be quite different from that of their colleagues in larger countries.

The peculiar nature of agricultural research in small countries suggests that information personnel should have an expanded role in the research system. The dependence on access to information in these countries instead of research and experimentation requires that information scanning, synthesis, and dissemination become an integral part of any research program. No longer can information personnel focus on acquiring, processing, and storing information material; they must move out and participate in the research work of their institute.

This expanded role will need support and commitments from management to ensure that the necessary changes in functions occur smoothly. More important, the skill mix of information professionals in each country will need to be reassessed and reevaluated to meet new needs and opportunities.

CHANGES IN THE ROLES OF INFORMATION STAFF

In a larger country, one might expect to have a library and documentation unit responsible for access to and storage of information used by researchers, a publication unit charged with disseminating the results of research through publications, and a training unit that organizes other events to inform clients of the latest research results and recommendations. In the small countries we studied, the functions of access, storage, and diffusion of information are combined. Trinidad and Tobago, the Gambia, Sierra Leone, Mauritius, Guyana, and Cape Verde all have core NARS institutions where the information unit in charge of the library also edits and produces the publications that document the work of the research service. In these countries, scientists have access to important, locally produced information through the library and to current information through their research networks.

In small countries, because of the limited size and resources of the agricultural research system, individuals have to carry out more than one function, and they may need training in order to be able to do this effectively. A librarian may need to be trained in editing and publishing, and an editor may need to be trained in the management of information.

In small countries, information managers face tremendous challenges in meeting the demands of small research systems. They have to wear several hats: as librarians, editors, and information specialists. They have to balance a complex assortment of demands for and sources of information. They need to work closely with research managers in order to determine the directions that demands for information are likely to take. And they must understand national policy issues so that they can keep policymakers well informed about agricultural issues.

LOCAL KNOWLEDGE IS A VITAL RESOURCE

Research systems in small countries have produced a great deal of knowledge that is not recorded. Scientists outside the country, and even scientists within the country, are often unaware of the valuable research work that takes place in the country. In some cases, experimental work is repeated unnecessarily because the results were not properly documented in the first instance or are not available. There is an abundance of aggregate data and information on global issues and problems but often a shortage of information on specific conditions and technological adaptations in the unique environments of individual countries. One of the most valuable contributions that researchers can make is to document the conditions and responses to new technologies in their own countries.

The need to document what is taking place within the research system is more than just good science; it is good policy and management as well. Policymakers may not be aware of research results. In our survey of research systems in small countries, it became apparent from the lack of available information that many systems have neither documented their research nor the direction in which it is going. Guinea-Bissau, Benin, and Sierra Leone are small countries that have recently begun to produce regular research bulletins that document the progress and achievements of research to a wider audience, both within and outside the country. It is quite likely that Sierra Leone's important research on cassava for use as a fresh vegetable is not well known in West Africa, or in Africa in general, but it can be shared by making the information available in local newsletters. In a similar vein, documenting the role of Guinea-Bissau's tiny research service in selecting technologies suitable for the country's farmers would gain it many friends in government and in the larger community of agricultural scientists.

In many smaller research systems, visits by consultants, foreign researchers, and technical experts are a ready source of much new information, which can also create some management problems. In exchange for providing expertise or information, visitors expect to receive information about how research in the country is organized and what the research system is doing. This takes time, which managers of small research organizations seldom have, nor do they have the staff to whom this function (what we call *science diplomacy*) can be delegated (Hobbs 1992). If there is a well-produced report on the research system and its institutions and activities that can be given to visitors, it can be a way to save time, gain friends, and demonstrate that the system is well structured and managed.

There is no doubt that access to and use of information from external sources is a key ingredient for successful and relevant research in small countries. What is often overlooked is that using scientific information and technologies developed *outside* the country depends upon reliable and accurate information on research and the state of agricultural resources *inside* the country. The better and more accurate the base of information on local needs and resources, the easier it will be to scan and select from the wealth of global scientific information and technologies. The information manager has a key function to encourage and organize the documentation of local research experience and knowledge.

Documenting and making local information on agriculture and natural resources available is also crucial for implementing "ecoregional approaches to natural resource management". Advances in information technology now make it possible for small countries to build and maintain local data bases, which can promote the use of standardized and comparable systems for the classification and storage of national-level data and information. Access to these data bases can be provided by the system's information services. And by linking these data bases with those of other countries, research in small countries can make an important contribution to global agricultural science.

Notes

1. The ideas and conclusions presented in this chapter are based on a companion ISNAR study on the management of scientific information, which was sponsored by the Technical Centre for Agricultural and Rural Cooperation (CTA) of the EEC/ACP, Lomé Convention. The studies and conclusions are available in a set of Small Country Study Papers authored and coauthored by P.G. Ballantyne and available from ISNAR or CTA.

9 Small Countries as Partners in Regional Initiatives

INTRODUCTION

In recent years, donor agencies, development banks, and international agricultural research centers (IARCs) have all lent their support to developing regional research in the belief that a research institution in one country can develop technologies useful to others (Box 9.1). Indeed, some research problems, such as control of pests and diseases and the management of certain natural resources, require a regional approach and pooling of resources to address them. Moreover, many donors now believe that developing technologies on a country-by-country basis is uneconomic, certainly in the smaller countries. This is the underlying rationale behind recent initiatives such as the "Regional Poles" and "Frameworks for Action" proposed by the Special Program for African Agricultural Research (SPAAR) and the numerous networks[1] supported by FAO[2] and IICA.[3]

Box 9.1. Why Take a Regional Approach?

- to exchange information and combine the collective experience of professionals in the same field
- to achieve economies of scale and efficiency by concentrating scarce human, financial, and other resources on key national and regional problems
- to minimize duplication
- to capture the effects of research spillover
- to rationalize human resource development
- to mobilize research efforts on transnational problems that require collaboration between countries

---

These external interests have placed increased pressure on national research institutions to devise their own plans and strategies for participation in regional activities. In small countries, however, research leaders and policymakers wonder whether their participation in regional activities might distort national research efforts to such an extent that national goals are not achieved (Gakale 1992).

In this chapter, we are concerned with how smaller NARS fit into regional systems and whether regional superstructures can deliver the research outputs and information that small countries are unable to produce on their own. Within each region there are

also international organizations and international centers that are increasingly working in partnership with national research systems. These partnerships may be more
difficult or less balanced when one of the partners is a country that can only afford
to maintain a small research system. This chapter reviews experiences with regional
approaches to agricultural research in four regions and seeks to answer the following
six critical questions that are faced by small-country NARS leaders:

- What functions should regional organizations assume in order to assist small-
  country NARS, and which functions must remain at the national level?
- Which of the different models of regional collaboration are most effective, and
  in which context?
- What are the preconditions for a cost-effective flow of knowledge and technol-
  ogy in a region?
- How should small-country NARS be organized to maximize the benefits of
  technology and information spillovers from neighboring countries?
- How can small countries influence regional research agendas?
- What are the inputs expected of small-country NARS in a regional system?

Since "regional approaches" can refer to a wide range of activities, it is useful to
define some key concepts. From a national point of view, a regional approach is a
national research strategy that promotes active participation in regional research
programs. It is also one that incorporates research conducted elsewhere in the region
into national planning and priority setting.

Three main approaches or mechanisms are employed. First, there are networking
efforts to promote regional cooperation in research (Beye 1992). Second, there are
regional organizations that promote research coordination and collaboration. Finally,
there are regional research institutes that conduct research themselves. Our main focus
in this chapter is on the latter two categories; networks are discussed in Chapter 10.

ORGANIZATIONAL CONTEXT FOR
REGIONAL COOPERATION

Many institutions have some involvement at the regional level in agricultural research
in developing countries. It is important to understand how they relate to each other
and what their various roles are. The structure of regional research in three of the
world's developing regions is shown in Figure 9.1.

At the highest policy-making level of the hierarchy are the continental or hemi-
spheric intergovernmental organizations, such as the Organization of African Unity
(OAU), the Organization of American States (OAS), and the Association of South-
East Asian Nations (ASEAN). The scope of these intergovernmental bodies is
open-ended, depending on the authority vested in them by member states.

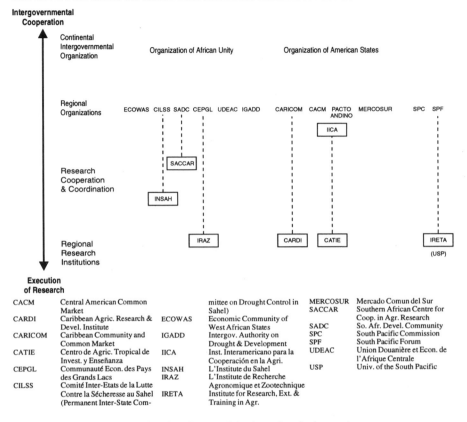

Figure 9.1. The structure of regional research in three developing regions

Operating at a level below this are (sub)regional organizations with specifically defined mandates (in trade, economics, the environment, agriculture, or development) to promote cooperation within a region. They include organizations such as the Caribbean Community (CARICOM), the Southern African Development Community (SADC), the Economic Community of West African States (ECOWAS), and the Communauté Economique des Pays des Grands Lacs (CEPGL), the latter three all in Africa. Belonging to a subregional community usually implies relinquishing some national powers to a central authority; it can also imply, as in Europe, belonging to a common market and a common currency area.

At the same level are various specialized authorities, such as the two authorities devoted to combating drought in Africa, the Comité Inter-Etats de la Lutte Contre la Sécheresse au Sahel (CILSS), and the Intergovernmental Authority on Drought and Development (IGADD) (see IGADD 1992a).

The importance of agriculture in the economies of member states has, in recent years, led these intergovernmental organizations to show increasing interest in this sector. In several instances they have created (sub)regional coordinating bodies for agricultural research and development. The Southern African Centre for Cooperation in Agricultural Research (SACCAR) and the Instituto Interamericano para la Cooperación en la Agricultura (IICA) are prime examples. In other cases, they have established regional research institutes to work on problems of regional significance and to complement the efforts of national programs: the Institut du Sahel (INSAH), created by CILSS; the Institut de Recherche Agronomique et Zootechnique (IRAZ), created by CEPGL; the Caribbean Agricultural Research and Development Institute (CARDI), created by CARICOM; and the Centro de Agricultura Tropical de Investigación y Enseñanza (CATIE), created by IICA.

These institutions operate at the fourth level of the hierarchy, below which come the various networks and collaborative programs they operate. Returning to the third level of the hierarchy, two regions with many island states have created commissions to promote regional cooperation and coordination: (1) The Indian Ocean Commission, representing Madagascar, Mauritius, Seychelles, Maldives, Comoros, and Reunion, and (2) the South Pacific Commission (SPC), serving 22 countries and territories, including the industrialized states of Australia and New Zealand along with the smaller Pacific island nations. The SPC also has its counterpart regional research institute in the Institute of Research, Extension and Training in Agriculture (IRETA).

One point often overlooked is that regional cooperation is an old idea (Box 9.2). Many of the national research institutions that small developing countries inherited at independence were originally regional in focus. In Sierra Leone, the Rice Research Station at Rokupr is a good example. For many years, tiny São Tomé e Príncipe had impressive cocoa research facilities that functioned as part of a global network on cocoa research, but which could hardly be justified to meet national needs. The adjustment in these institutes from a regional perspective to something much smaller has been difficult and has contributed to the problems of many small countries.

REGIONAL APPROACHES TO AGRICULTURAL RESEARCH IN THE CARIBBEAN

Agricultural research in the Caribbean has been organized on a regional basis for two decades. A regional approach was recognized as a necessity from the outset, since many of the region's tiny countries could not afford national programs of their own. The key to success is a division of labor between national programs and the regional research institute, which is clearly spelled out in a regional policy. As a result, national research capacity and scope cannot be assessed without considering the regional context.

Through CARICOM,[4] the high-level intergovernmental body, policymakers and research leaders define regional policies to attack common problems and, at the same

---

Box 9.2. Historical Perspective on Regional Cooperation in Africa

In Africa, regional research existed before independence in each of the colonial subregions. Interactions between them was effected through mechanisms like the Commission de Coopération Technique en Afrique au Sud du Sahara (CCTA) and the Conseil Scientifique pour l'Afrique Sud du Sahara (CSA) at the regional level, while agencies such as the Comité Régional de l'Afrique Centrale pour la Conservation et l'Utilisation des Sols de l'Afrique Meridionale operated at the subregional level.

Anglophone Africa had a number of interterritorial research institutes such as the six institutions under the West African Agricultural Research Organization, which included the West African Cocoa Research Institute, The Rice Research Station at Rokupr, and the West African Institute for Oil Palm Research. The Empire Cotton Research Organization covered all countries where cotton was a major crop. The East African Agriculture and Forestry Research Organization, the East African Veterinary Research Organization, the East African Fresh Water Fisheries Research Organization, and the East African Marine Fisheries Research Organization were responsible for subregional research activities in Kenya, Tanzania, and Uganda.

In francophone and lusophone Africa the situation was similar. Research in the French territories was carried out under the Office de la Recherche Scientifique et Technique d'Outre-Mer (ORSTOM) and the commodity and agronomic research institutes that are now part of the Centre de Coopération Internationale en Recherche Agronomique pour le Développement (CIRAD). ORSTOM was, and remains, oriented toward basic research. Belgium had one research organization for all its territories, the Institut National pour l'Etude Agronomique du Congo (INEAC), which served all of its territories. The Portuguese had interterritorial research institutes and research units for their dependent territories in Africa under the Junta de Investigaçãos Científica do Ultramar (JICU). The *junta* also maintained tropical agronomic and veterinary institutes in Angola, São Tomé, and Mozambique and agronomic missions in all its five territories, including Cape Verde and Guinea-Bissau.

Since independence, each country has tended to develop or maintain its own resources. Even where experience has shown the benefits and returns to joint research investments, there has been little enthusiasm for continued support to regional research from the formerly cooperating countries. As sovereign states, overall authority and common policies across countries could no longer be enforced; therefore, it has become necessary to develop new structures and voluntary mechanisms for intercounty regional cooperation.

Source: Beye (1992).

---

time, delineate those areas that are of purely national concern. This same policy body allocates resources to CARDI, the regional research institute. For many of the smaller island states, CARDI is the principal executor of their research. It therefore has a special obligation to the smaller countries. In some of the larger countries with formal research institutions, responsibilities for national research are shared among national and regional programs.

The current allocation of regional and national responsibilities has emerged slowly, following the evolution of regional planning and policy-making bodies. CARICOM, created in 1973 by the Treaty of Chaguaramas, formulated a regional food plan in

1975. This plan was expanded to include health and education and a regional food and nutrition strategy in 1980. Other policy initiatives that followed during the 1980s included the development of a common market, strategies for agricultural diversification and exports, the launching of programs on agroforestry and fisheries, and the setting up of advisory services for natural resource use and conservation.

CARDI is only one of several new institutions founded to implement these plans. The Caribbean Food Corporation (CFC) was formed to promote and finance agribusiness. The University of the West Indies (UWI) was established to meet regional training needs. A regional export development program, the Caribbean Export Development Corporation (CEDC), has also been launched, together with an Agricultural Diversification Coordinating Unit (ADCU).

The Standing Committee of Ministers responsible for Agriculture (SCMA) now oversees all public-sector agricultural research in the region. It also serves as CARDI's Board of Directors, seeks resources from donor agencies, and allocates funds to regional agricultural projects. It is assisted by the CARICOM Secretariat and advised by national agricultural planners, who also meet as a regional body. Bilateral assistance does not, however, come under the purview of the SCMA, nor do the universities, which remain relatively free to formulate their own programs.

REGIONAL SCOPE OF RESEARCH

The most striking feature of regional research in the Caribbean is the great diversity of commodities covered (Table 9.1), made possible by the allocation of responsibilities to a wide range of actors. Research on the major global staples is handled mainly through networking arrangements with the international agricultural research centers, with a great deal of technology imported from outside the region. Most of the research at both regional and national level is adaptive in nature, restricted to the testing of existing technologies.

This leaves room for applied research on a wide range of minor food crops, including breadfruit, garlic, onions, hot peppers, and edible aroids (eddoes, dasheen, and tannia, varieties of *Colocasia* and *Xanthosoma*). These crops are important to the region's food security and nutrition but are not well covered by international research efforts. Hence, they are the focus of both applied and adaptive research by CARDI.

Traditional export crops such as sugar and tobacco are well covered by a few strong programs in small countries. Research on nontraditional exports is conducted largely by private industry but with support from CARDI where the venture is considered too risky for private enterprise working alone. Thus, CARDI does research on broccoli. It also works jointly with Jamaica and Trinidad on citrus; with Jamaica, Trinidad, Dominica, St. Lucia, Antigua, and St. Vincent on passion fruit, mango, pineapple, and soursop; and it works on flowers and ornamentals in Trinidad, Barbados, and Jamaica, where it shares a project with the University of the West Indies. Most of the resources go to crops; apart from some applied research to improve

Table 9.1. Domains of the Regional Research Portfolio in the CARICOM Caribbean

Global Staples	Trad. Exp. Crops	Minor Food Crops	High-val., Nontrad. Exp. Crops	Livestock	Socioecon. & Rural Engin.	NRM
Beans	Bananas	Breadfruit	Broccoli	Small	Farm manage-	Land use &
Cassava	Cocoa	Garlic	Citrus (limes,	ruminants:	ment	water mgmt.
Groundnuts	Coconuts	Onions	grapefruit)	(goats, sheep)	Farming	Soil (fertility,
Potatoes	Coffee	Peppers	Flowers		systems	erosion,
Rice	Cotton	Pigeon peas	Ornamentals	Cattle	research	conservation)
Soya	Oil palm	Plantain	Mangoes		Postharvest &	Water mgmt.
	Sugar	Sweet potatoes	Passion fruit	Swine	storage	Range &
	Tobacco	Eddoes, Tan-	Peaches		Machinery &	pastures
		nia, dasheen	Pineapples	Feeds &	tools	
		(*Xanthosoma,*	Soursop	nutrition	Agro-	
		Colocasia)		Animal	processing	
		Yams		breeding		
		(*Dioscorea*)				

local breeds of cattle and goats, there is less research on livestock than one would expect, given the high levels of imports and the successes that research has had in the past (Parasram 1992).

MANAGING THE REGIONAL RESEARCH SYSTEM

In summary, the regional research system in the Caribbean is characterized by maturity and diversity in the institutions responsible for research and development, allowing complex and multifaceted relationships among the various actors. This said, it is not easy to manage all these actors, to make sure each one plays its part and performs its role. But the full cast has been assembled.

The Caribbean is noted for its strong national and regional commodity organizations. At the regional level there are parastatal research organizations, such as the West Indies Sugar Cane Breeding Station (WICBS), and for bananas there is the research and development division of the Windward Islands Banana Growers' Association (WINBAN) based in St. Lucia. These organizations derive much of their funding from the industry itself, through a tax on exports. They are reputed to be highly responsive to the needs of their users, both producers and consumers. It is interesting to note that there is also a small regional research facility in cocoa based at the University of the West Indies.

Many of the countries in the region have some national research capacity attached to commodity boards and parastatals for their traditional exports. Some national commodity boards, such as the coffee and citrus boards in Jamaica, contract their research out to CARDI or UWI rather than conducting it themselves. Still others, such as the Jamaican Coconut Industry Board, conduct their own research but seek collaboration with other agencies in certain areas of strategic research. It is envisaged that as the region diversifies further, commodity-oriented research organizations may become still more important, especially within the parastatal and private sectors.

There are consolidated research departments and institutes within ministries in a few of the larger countries such as Guyana, Trinidad and Tobago, and Jamaica. Guyana is the only country with a consolidated national agricultural research institute (Forde 1992).

The problem of coordinating public research at both the regional and national levels, as well as coordinating research across countries with differing capacities and demands, is a difficult one. The mechanisms are in place within the ministries and the CARICOM bodies that establish regional agricultural research and development policies, and this works reasonably well for public institutions at both the national and regional levels. However, more work is needed to establish a forum for coordinating public, parastatal, and private research entities in a regional system.

REGIONAL APPROACHES TO AGRICULTURAL RESEARCH IN SOUTHERN AFRICA

For many years, the political problems associated with the Republic of South Africa have forced the region's front-line countries to work together. The presence of an external threat, *apartheid,* was a major factor in the foundation of SADCC,[5] the forerunner of today's Southern African Development Community, SADC. However, as changes sweep the Republic of South Africa and *apartheid* no longer poses a major threat to the stability and welfare of neighboring states, the rationale for regional collaboration has not disappeared. On the contrary, the transformation of SADCC into a community symbolizes the increased level and scope of regional cooperation. The political urge to collaborate, combined with the support of many development agencies and donors, means that the region now probably has the developing world's most comprehensive mechanisms for cooperation in agricultural research.

SACCAR was created in 1985 to strengthen agricultural research and training by promoting two regional functions. One is to maximize the use of scarce resources through more intensive collaboration among members. The other is to provide a mechanism to improve the access and management of links to external sources of knowledge and technology.

Unlike CARDI, SACCAR is not a research institute. It serves as a mechanism for discussion, a facility for information exchange, a source of training and education fellowships, and a catalyst in the relationships between national agricultural research systems, international agricultural research centers, and donors. SACCAR is an important counterweight for small national research groups in their relationships with larger systems and with external agencies. This role is explicit in its overall objectives (Abe and Marcotte 1989), which state that SACCAR should provide a forum in which small countries can participate as equal partners with their larger neighbors (Box 9.3).

SACCAR's main modes of operation are a series of regional research programs and projects. Examples include the sorghum and millet improvement program based in

---

Box 9.3. The Role of SACCAR

For small countries, SACCAR

- provides a forum in which small NARS can participate as equal partners with their larger counterparts;
- works with the NARS to generate improved technologies for high-priority commodities and factors of production; (This includes assistance to NARS in the planning, execution, monitoring, and evaluation of their research, their training in research management for national scientists, and the development of specializations in key areas that enable the whole region to benefit from national outputs in training and research.)
- provides long-term mechanisms for collaborative research and training, linking individual NARS to other NARS of the region as well as international research agencies and donors; (This implies the establishment of coordinated regional research programs as well as a mechanism for formulating joint national priorities.)
- provides training for national scientists that allows them to actively participate in the global research system and to build up the regional scientific community;
- disseminates information through publications, provides a monitoring service (an early warning service) in strategic areas relating to food security, and assists NARS in strengthening their own ability to organize and manage information resources.

Source: Okello and Eyzaguirre (1992)

Zimbabwe, the grain legume improvement program based in Malawi, Mozambique, and Tanzania, and the land and water management programs based in Botswana. In addition, there are regional projects such as the research management training project, the agroforestry research project, and the regional genebank. These programs and projects are usually executed by external agencies in conjunction with national systems, and a number of support services have been set up, including an information service, research data bases, regular meetings and workshops, reports and publications, and a sponsored regional journal. These provide an organizational base for many research networks, which is also a means for directing the flow of resources from IARCs, donors, and international organizations to the research programs of member countries.

The first generation of regional research projects in Southern Africa was planned and implemented almost exclusively by IARCs. Through SACCAR, national systems have had a great deal more say in the planning phase, as well as an expanded role in implementation and evaluation. National scientists have also been involved through participation in feasibility studies and regional planning workshops. During implementation, mechanisms such as steering committees, technical advisory panels, and annual research workshops have ensured that project activities are relevant to regional needs.

Most regional networks seek to involve national research groups from each of the 10 SADC member countries. The way this works is that a particular country assumes responsibility for a specific commodity. Malawi, for example, has taken the lead in groundnut research, and Zimbabwe, in maize, while Botswana and Lesotho have concentrated on beans and cowpeas, respectively. Although individual countries take the lead, other countries must maintain research activities in each crop. Commodity research is a condition for membership in the regional commodity programs. The opportunities for a small country to discontinue its research on a specific crop and to rely on its neighbors for new technologies are therefore limited, since the country would then be excluded from the network on the grounds of nonparticipation. And technologies would be difficult to locate and borrow without the exchange mechanisms offered by the network. SACCAR's networks therefore tend to further stretch the already broad agenda of smaller national systems. There is thus a contradiction between SACCAR's stated objective of benefiting smaller countries and the way its networks operate, which in effect places further demands on the countries.

All SACCAR's regional programs are implemented at the national level and are managed by the national research groups themselves. This means that it is sometimes difficult to distinguish between national and regional research activities. There is also a tendency for national programs to duplicate or mimic regional programs in the hope of attracting external funding away from them. Besides creating unnecessary competition and conflict, this can lead to distortion of national research priorities and to uncertainty as to what should be done nationally or regionally.

The problem of managing regional activities within national systems can be acute for the smaller systems. While their larger neighbors have benefited because their scientists were able to collaborate as equal partners with international scientists, smaller systems have been disadvantaged, lacking experienced researchers who are able to identify the technologies relevant to national needs. As a result, the staff of smaller systems have tended to serve as technicians for the multilocational trials of international scientists. Thus, although national programs have been able to influence the research direction of some IARCs in the region, the smaller NARS have difficulties implementing regional projects as equal partners, despite the interventions of SACCAR.

REGIONAL APPROACHES TO AGRICULTURAL RESEARCH IN WEST AFRICA AND THE SAHEL

All the region's countries, both large and small, face similar agricultural challenges. Agriculture is the major employer and largest contributor to gross domestic product in all countries in West Africa and the Sahel. Rural poverty, accompanied by stagnant or declining agricultural productivity, remains a major problem throughout the region, but it is exacerbated in the smaller countries, where annual per capita agricultural income was less than US$150 in 1990. The resource base for agriculture is fragile, with rainfall declining over the past 25 years; areas with greater rainfall have considerable problems managing resources because of declining soil fertility as pressure on the land rises. The gap between potential and actual yields is very high, and the availability of innovations attractive to farmers is limited. Such traditional exports as tropical beverages, oilcrops, and cotton have been severely affected by the declining terms of trade, which have hit the small exporting countries particularly hard. The potential for developing high-value nontraditional exports is also frustrated by the lack of any transport, marketing, or postharvest infrastructure.

Small West African and Sahelian countries have tended not to think of themselves as a block on virtually any issue. They are separated from one another by geography, language, colonial tradition, ethnicity, and political system, and these barriers have proved difficult to surmount. Among their closest neighbors, they all have larger countries with bigger research institutions, and in general have stronger links with these countries than with each other. The region has two large blocs: the francophone countries in a single monetary zone linked to the French franc and the anglophone Commonwealth countries. Outside these blocs are the two lusophone countries of Cape Verde and Guinea-Bissau (Gilbert, Matlon and Eyzaguirre 1993).

National agricultural research systems in the region are generally of recent origin, having been created since independence. They remain weak in terms of both trained scientific manpower and research management skills, especially in the small countries (Eicher 1988). In addition, many West African governments have suffered severe financial crises affecting their ability to fund research. In the current climate, national research systems are under pressure to generate useful technology and demonstrate impact in order to justify their claims for scarce public funds. Failure to demonstrate this impact could plunge the region's public-sector research into deep crisis.

For small countries, the research institutions are particularly vulnerable to collapse; some research institutes are now held together by only a handful of key senior staff. As capacities and performance levels drop below the level necessary to sustain institutional life, there may be calls to downgrade these institutes to performing adaptive research or acting as liaison units within development agencies, along with consolidating research into regional centers and organizations that can maintain sufficient critical mass and funding.

Many attempts have been made in this region to foster regional collaboration. However, no single regional body has been able to coordinate policies on agricultural development and, hence, to promote common research efforts. Strong divisions along linguistic lines, divergent political structures, and even the drastically divergent scales of the NARS themselves hinder regional collaboration. Although mechanisms to exchange information and coordinate research abound in response to similarities in agroecological conditions and the pressing nature of common problems, these often overlap and compete with national objectives and regional aims rather than complementing them.

While the outreach activities of CGIAR centers and other regional bodies and programs, such as ITC, CEAO, and SAFGRAD, contribute to regional collaboration, only two formal institutional mechanisms for regional collaboration exist: the Institut du Sahel (INSAH) (see Box 9.4) and, more recently, the Conference des Responsables de la Recherche Agronomique Africains (CORAF). The underlying principle behind

---

Box 9.4. The Institut du Sahel

Following the Sahel drought of 1968–73, it was realized that agricultural research had a major role in the long-term rehabilitation and development of the subregion. The Institut du Sahel (INSAH) was created in December 1977 and became operational in 1978.

The institute was created to implement regional cooperation in the areas of agro-sylvo-pastoral research, training, and dissemination of scientific and technical information in the subregion. Its permanent mandate is to

- collect, analyze, and disseminate the results of scientific research
- transfer and adapt appropriate technologies
- coordinate, promote, and harmonize scientific and technical research
- train researchers and technicians
- reflect upon and define regional research themes
- plan research at the regional level

Research programs coordinated by INSAH focus on varietal improvement, soil and water management, drought resistance, and food security. These are supported by a regional documentation network (RESADOC). Efforts are currently in progress to broaden the field and scope of intervention to include important research areas such as animal production (especially small ruminants) and natural resources.

While no specific program has been developed for small Sahelian NARS, INSAH seeks to address their problems in a variety of ways. First, a Travel Grants Program enables researchers to benefit from networking trips to other NARS whose research programs are of particular interest. For example, researchers from Guinea-Bissau visited the socioeconomic research unit of ISRA in Senegal, and the horticultural program in the Gambia made contacts with the horticulture and fruit section of IER in Mali. Second, INSAH supports requests for technical assistance from small NARS. Third, within the context of collaborative programs, opportunities are created for scientists of small NARS to benefit from training experiences within larger NARS. Increased attention is now being given at INSAH on how best to develop adequate mechanisms at both the national and regional levels to promote more vigorous publication and diffusion of research results.

Source: Jallow (1992).

---

these efforts is to create complementarity among national research programs, which appears both logical and necessary. It is, perhaps, best expressed through participation in regional programs that are mutually beneficial.

The weaknesses of many Sahelian NARS motivated INSAH to elaborate a five-year strategy and program (1990–1994) and, more recently, in collaboration with SPAAR, to develop a framework for revitalizing agricultural research in the Sahel (INSAH/SPAAR 1991). The principal elements of these two complementary initiatives are

- *Institutional reforms of NARS:* The objective of the proposed reforms is to create an enabling environment that promotes creativity and innovation and rewards performance. The reforms will reinforce the position of the NARS as the basic building block for a sustainable regional research system. Solutions to the fundamental size, capacity, and stability issues confronting NARS are proposed; reforms will help NARS set priorities in a regional context and define new mechanisms for greater financial stability and management autonomy.

- *New mechanisms of regional collaboration:* Where regional research priorities are clear, resources will be consolidated around key institutions or regional research poles (lead national centers) in the region (INSAH 1990). The poles will concentrate on developing a critical mass of well-trained and motivated scientists, adequate resources, and efficient and flexible management systems. By virtue of this, the poles will be the main focus for rapid advances in technology generation. These lead national centers are expected to find solutions to the problems of small NARS and also to serve as centers for training.

One problem that remains outstanding is the institutional context that can support regional research planning and execution. There are many competing, overlapping, and independent regional initiatives and frameworks, which in turn seldom coincide with the agroecological regions defined for the purposes of technology generation and transfer and natural resource management. Are new organizational mechanisms needed? Or can existing regional institutions work together to assume a greater role in research policy and funding?

While the need for regional collaboration in West Africa and the Sahel is undeniable, the usefulness of a regional organization as a mechanism for funding national research is debatable. One problem with such organizations is that they come to be perceived as primarily a mechanism of donor convenience. The verdict is still open on whether such organizations can really promote the more efficient use of resources within a region, or whether they are just one more organization with privileged access to donor funding.

THE IARCS AS REGIONAL PARTNERS

There are two international agricultural research centers based in West Africa: the International Institute of Tropical Agriculture (IITA) in Nigeria (Box 9.5) and the West African Rice Development Association (WARDA) in Côte d'Ivoire (Box 9.6). A third, the International Crops Research Institute for the Semi-Arid Tropics (ICRISAT), has a major station in Niger to serve the Sahelian countries. These centers provide expertise, technologies for testing and adaptation, operating funds for conducting experiments, and a venue for meetings and training to share knowledge and overcome the isolation that national scientists may experience working within a small-country NARS.

Although IARCs have operated in Africa for more than 20 years, their relationships with NARS remain ambivalent. With the exception of human resource development, it is generally recognized that they have had only limited success in strengthening NARS.[6] There are several reasons for this. First, international centers have acted as central sources of technology for the regions in which they are located. This may have

---

Box 9.5. The Research-Liaison Scientist Scheme of IITA

If NARS are to rely on international centers to do applied research on problems that constrain national production, better links are needed to transmit NARS needs to the centers. The centers will in turn need to acquire a better understanding of the varied resource endowments and ecological conditions in the region.

In response to this demand, IITA has created a linkage component, the research-liaison scientist within selected NARS. For some of the smaller countries, the liaison scheme provides assistance in determining the technical assistance requirements of the NARS and preparing the necessary funding proposals. In some cases, this involves identifying particularly "vulnerable" NARS for accelerated training assistance. This is done by means of resident teams of research-liaison scientists who assist the NARS in adopting and using a systems orientation in research on crops and resource management. The liaison scientists are expected to spend 25% to 30% of their time on a research concern of the host NARS. This type of program is consistent with the overall strategy that is emerging for small countries. It is notable, however, that Benin is still the only small country to benefit from this approach. Nonetheless, the concept of a research-liaison scientist is a useful model for researchers in small countries who must often work in this mode.

Source: Suh et al. (1992).

Box 9.6. WARDA's Research Partnerships for Rice in West Africa

WARDA's collaboration with national research groups operates through regional task forces, each composed of all the national rice scientists in the region who are working on a given problem or theme. The task forces fulfil four general functions: (1) joint research planning, (2) technology transfer, (3) dissemination of information, and (4) allocation of assistance for regional research activities. Each task force has developed a regional master plan to guide task sharing. To develop this plan, members began by identifying the priority constraints to rice production within their area of specialization. After reviewing past research, regional research priorities were established and the relative strengths and weaknesses of each national system in each priority area were critically examined. Finally, specific tasks were allocated to certain national programs on the basis of national comparative advantages. Programs with strengths in particular disciplines were identified to play lead roles, working with WARDA in advanced research to generate new technologies for the benefit of the entire region. All national programs participate in the testing of new technologies and in the exchange and dissemination of research results. With access to special project funding, the task forces provide small grants to assist participating national programs. Both larger and smaller national systems benefit from these grants.

Source: Gilbert, Matlon and Eyzaguirre (1993).

---

led, inadvertently, to the centers imposing their own priorities and workplans onto those NARS wishing to collaborate. The result has been a strict stratification in roles, with IARCs often claiming the responsibility for basic and applied research aimed at technology generation and leaving NARS the secondary role of adapting and testing technologies coming from the international centers.

Second, IARCs have often failed to build upon the institutional diversity present among national programs. They have either failed to distinguish among NARS or they have tended to concentrate their collaborative efforts primarily in the larger NARS.

Third, IARCs tend to collaborate with NARS on the basis of projects rather than programs. The short time horizons imposed by donors put pressure on the IARCs to produce quick results, generally through technology transfer, with insufficient efforts directed at building NARS capacity. The project nature of collaboration has also tended to fragment NARS internal activities into distinct research components. In its worst form, one can observe IARCs competing for the scarce time and resources of national scientists in separate collaborative projects. The sustainable benefits of such activities are few, with most suffering from what one might call "post-project collapse syndrome".

Finally, many IARC initiatives with NARS have been designed with a bilateral focus in which needs for new technologies and research support are assessed on a country-by-country basis without a broader regional analysis of the relative research capacities and needs among all the NARS in the region. Most IARC networks established in Africa tend to be "central-source" networks, with policies, direction, and flows of information and technology emanating from the IARC at the center to participating NARS at the periphery.

One consequence of this has been a general failure to establish working linkages between NARS themselves and, thus, the failure to develop and exploit potentially significant research complementarities between national programs. Because of planning in isolation and a lack of information on the needs, resources, and activities of neighboring NARS, there is widespread duplication of efforts, while opportunities for complementary research efforts and technical spillover are rarely exploited.

Some IARCs have responded by proposing a radical restructuring of NARS–IARC relationships, based on jointly developed partnerships. An important new aspect of this approach is the view that all research participants, whether they are within NARS or IARCs, are part of an integrated and interdependent regional system. The partnerships have two basic objectives. The first is to achieve more complementary and efficient sharing of research tasks among NARS themselves as well as between NARS and IARCs. This would be done by allocating responsibilities on the basis of comparative advantage. The second objective is to achieve scientific critical mass on a regional scale. Both these objectives carry far-reaching consequences for the way research planning is conducted.

The WARDA experience suggests that in planning research with a regional perspective, it is not appropriate to generalize different roles on the basis of the size of national systems alone, nor is it safe to assume that the strength or weakness of a national system is uniform across crops or disciplines. Most national systems are characterized by considerable heterogeneity across research themes and over time because of staff turnover, funding crises, and other factors. Rather than a uniform gradient of weaker to stronger national systems, we observe an uneven mosaic in which well-trained scientists are carrying out excellent research in the midst of an otherwise weak program. A pragmatic and flexible approach is necessary to identify and fully involve the many islands of excellence that exist within the NARS and to create an enabling atmosphere for national scientists to fully exploit their capacities.

One of the most common complaints of researchers in West and Sahelian Africa is that they feel isolated. Scientists in small national systems often lack adequate knowledge of the programs and research results of neighboring systems and regional centers. This sense of isolation is aggravated by poor information services, particularly publishing, at both the national and regional level, as well as by differences in language and poor communication between countries. In West Africa, fragmentation is largely an inheritance from the colonial era. It is to be hoped that the numerous external agencies active in the area can work to restore the links that unify the region's countries through common ecological zones and commodities.

In summary, regional collaboration in West Africa has proved elusive. No single subregional body has been able to coordinate policy on agricultural development and hence promote a common research effort. In response to similarities in agroecological conditions and the pressing nature of common problems, mechanisms for exchanging information and, supposedly, coordinating research abound, but these often overlap and compete with national objectives rather than complementing them. Strong divisions along linguistic lines, divergent political structures, and the very different scales of national systems continue to hinder collaboration.

REGIONAL APPROACHES TO AGRICULTURAL RESEARCH IN THE SOUTH PACIFIC

Many of the South Pacific island nations are among the smallest countries in the world. And even in the days before independence, when many of these countries were UN trusteeships and colonial territories, it was assumed that agricultural development in the South Pacific would be undertaken regionally. There are enough similarities within the ecologies and food production systems of Melanesian and Polynesian cultures to make a strong case for regional collaboration, but efforts to define a regional research agenda have so far been piecemeal and largely unsuccessful.

The size constraint in this region is very pronounced.[7] Only Fiji and Papua New Guinea have research organizations with more than 20 scientists, and nine of the region's countries have virtually no national research capacity at all. Compounding the effects of the size constraint are the great distances that separate countries, incurring high costs in communication and in transporting produce to market.

An unusual feature of the Pacific region is that it includes developed countries such as Australia and New Zealand, as well as developing countries. There are also several territories, such as New Caledonia, American Samoa, and French Polynesia, whose research and development institutions are funded and managed as part of the larger scientific and technical research systems of the United States and France. These powerful developed-country actors provide many opportunities for technological development. At the same time, however, there is a tendency for them to dominate regional and national research agendas, preventing small, fragile national systems from articulating and meeting their real priorities. It also leads to parallel and, at times, competing regional initiatives.

Externally funded donor projects are particularly important in the Pacific countries. In several countries, almost all of the research is done by donor projects, with little input from the national system in establishing policies and priorities. Even in countries where there is some formally organized national research, donor projects tend to steer national activities towards the donor's priorities and objectives; there is relatively little analysis on whether donor-financed activities contribute to national goals. In Fiji and Papua New Guinea, where there is more substantial research capacity, there is a stronger basis for negotiation on national priorities and objectives with donor agencies.

There are two main regional organizations involved in research. One is the Institute of Research, Extension, and Training in Agriculture (IRETA), based at the University of the South Pacific (USP) in Western Samoa; the other is the South Pacific Commission (SPC), based in New Caledonia, French Oceania. IRETA conducts research primarily on staple foods, including taro, cassava, and sweet potatoes, as well as on farming systems and soil fertility and conservation. It draws heavily on the resources of the USP's Faculty of Agriculture and its extensive regional network. Another regional organization, the South Pacific Forum is particularly active in trade and transport issues.

Founded in 1947, the SPC is responsible for coordinating agricultural research in 22 countries, most of them small. It conducts no research itself but is nonetheless a powerful presence in the region, with a budget of US$ 26 million in 1992. However, it appears to rely mainly on expatriate management and allegedly provides research managers from national systems with few opportunities to influence regional policies and priorities. Its critics claim that unsuitable assessment methods lead the Commission to follow its own agenda in setting regional research priorities rather than responding directly to national interests and needs. As a result, it appears to compete with national programs, not only for human resources but also for grants and aid.

The SPC coordinates regional programs and projects in the areas of plant and animal quarantine, plant health, and plant protection. It has also supported marine fisheries research and has become increasingly active in the management of information for research and in issues involving natural resource management. The viability of a regional approach in the Pacific depends on the ability of the region's nations to make regional research programs self-sustaining, something they have not proved very good at so far.

The way forward is difficult to perceive. With the national capacities of the small island states at such a universally low level, it is difficult for a strong indigenous body to articulate regional policies and priorities without reflecting donor interests. The starting point will be for national systems to develop clear policies and plans from which a regional agenda can be determined. At this point, a single regional mechanism could come into play, formulating regional policy and coordinating research efforts. This could be a reformed SPC. CARICOM and CARDI appear to be a good model for both the SPC and IRETA to follow, but as matters stand at present, the Pacific countries have a considerable way to go in linking regional policies with regional and national systems.

TECHNOLOGY GRADIENTS

Linking research institutions and programs in a region to create economies of scale and scope presumes that there is a *technology gradient* in the region. We employ the term *gradient* because it implies that in a region with similar production systems, agricultural technologies and information can flow from countries where the use and generation of technologies is more intensive to those countries where it is less intensive.

To achieve a manageable gradient among national systems, a regional mechanism may be needed to identify the sources and facilitate the flow of useful technologies in a region. The proposed strategy would be to structure and orient national research to maximize benefits from the technology gradients that are identified. There are several factors that determine the gradient and govern the flow of specific technologies (Box 9.7). The first is the intensity of the technology-generating institutions within the region. These may be national or international in character, public or private. Their research focus may be general or confined to a single commodity or topic. And they may be, and often are, located in larger countries where there is greater research capacity and stronger demand for new technological inputs.

Where a technology gradient is very steep, we may find that there are few factors that impede the flow. For example, the availability and use of agricultural technologies in the Republic of South Africa is much higher than in the small countries that

---

Box 9.7. Features of a Technology Gradient

Conditions that define a technology gradient	Factors that affect the flow of technology and information
• the presence of institutions that generate agricultural technology among countries of the region with differing intensities of technology generation and use • the kinds of agricultural technology being generated and their global and regional distribution • similarities in the agroecological conditions and agrarian structure of the countries involved	• similar institutional arrangements and traditions in the organization of national institutions of science and technology, including the NARS • common language and similar sociocultural environments • political compatibility, especially among countries in a region

surround it. Despite the earlier political cleavages and lack of institutional framework for collaboration in agricultural research and development, a large number of new technologies were and still are informally introduced from the Republic of South Africa into the smaller SADC countries. However, because the socioeconomic conditions of agriculture in South Africa are markedly different from those in the SADC countries, NARS in these countries must still monitor the flow of new technologies and minimize those that could be inappropriate and wasteful with grave consequences for the long-term productivity of the resource base (see ISNAR 1989a).

A similar situation exists in the Pacific where there are important bases for technology generation in Australia, New Zealand, Hawaii, and the French Pacific territories. There is frequent travel by islanders to the developed countries of New Zealand and Australia and other territories, as well as consultants and visitors bringing and exchanging technologies. However, there is still a problem of adapting these technologies to the subsistence farming systems of most of the Pacific island nations.

Establishing a regional technology gradient requires policy and management mechanisms at both the national and regional levels. To say that there is technology available within the region is not enough. It must be selected, transferred, and adapted. Mechanisms are needed to match the sources to the needs. Even practical considerations of quarantine and national policies on input use, pricing, intensification, and diversification may need to be harmonized.

Where coherent, recognized regional forums exist, the prospects for establishing and managing a regional technology gradient are promising. However, competing regional mechanisms are not necessarily useful, as the cases of West Africa and the South Pacific illustrate.

COMPARING REGIONAL APPROACHES

Southern Africa and the Caribbean present two contrasting but largely successful models of institutionalized regional research. In Southern Africa the approach is built around a single mechanism, SACCAR, which promotes greater collaboration and division of labor among national research institutes. It also coordinates their joint activities with external agencies, including donors and international centers. The Caribbean's approach involves two institutes, one (CARICOM) in a high-level policy-making and coordinating role. The other is CARICOM's regional research institute (CARDI). This approach involves a division of labor between research of regional significance and that which is of purely national concern. Besides having collaborative programs with national institutes in the larger countries, CARDI meets the research needs of member states that are too small to maintain their own institutions.

The fact that both models are relatively successful suggests that success has more to do with factors other than the organizational structure selected. The most obvious common factor in the two regions is the predominance of a single language and shared institutional traditions. In both cases, this was a major factor behind the creation of a regional forum for policy-making and research coordination. Other common factors among Southern African countries include similar agroecological zones and the same food staples. However, a comparison with West Africa and the Sahel shows that these are not enough to establish a flow of technologies within a region (see Box 9.8).

Box 9.8. Achieving Successful Regional Collaboration in Research

Successful regional collaboration in research is easier to achieve when the following factors are present:

- political will to collaborate between the different countries of the region
- a common language and shared institutional culture
- compatible institutional structures for research and development
- common problems facing agricultural research and development, especially common crops and agroecological zones
- geographic proximity and good channels of communication
- institutional mechanisms at the regional level to establish goals, coordinate tasks, and manage common resources
- complementary research capacity and strength in the different sectors

---

One would assume that because of agroecological similarities, technology would flow easily from northern Nigeria, with its long-established research capacity, to similar environments in Niger and Chad; however, this has seldom been the case. Ecological similarities are often less important in the flow and transfer of technologies than institutional factors such as language and historical ties between institutes. This is why research results from northern Nigeria might flow more readily to the Gambia and Sierra Leone, where the research organizations share historical links and similar institutional cultures. As ecoregional approaches to research assume greater importance, institutional issues and forging new ties between institutions have emerged as a central issue (Goldsworthy, Eyzaguirre, and Duiker 1995).

In contrast to the coherent approaches in Southern Africa and the Caribbean, West Africa and the Sahel present a tangle of overlapping and conflicting regional policy frameworks, research coordination schemes, and other policy initiatives, each limiting the other's chances of success. The division of the region along monetary, institutional, and linguistic lines, inherited from the colonial era, is a continuing source of frustration and isolation. The search for a coherent mechanism for regional coordination at the policy level is, however, quickening. This bodes well for future opportunities to make the most of the research capacity within individual countries that is often confined to those countries and undervalued as well.

An important feature affecting the model of regional cooperation selected is the size of the national systems in a given region. In the South Pacific and the Caribbean, the two regions with regional organizations that conduct research, there are no large developing countries. Pooling institutional resources across countries is a prerequisite for national and regional development in these two regions. Both have a regional university, which physically hosts the regional institute as well as providing it with expertise.[8]

In contrast, Southern Africa and West Africa both contain several larger countries (notably Nigeria and Zimbabwe) with the capacity to generate technologies and information of regional significance. The opportunities for technology transfer are therefore greater within these regions. Southern Africa presents by far the more advanced prospect, with SACCAR beginning to make strides in capturing and directing an increased flow of technology and information.

If further initiatives at the policy level can establish a coordinating body similar to SACCAR in West Africa and the Sahel, a regional research institute may well prove to be unnecessary. The Institut du Sahel, created to encourage research on natural resource issues affecting agricultural development in the Sahel, has wisely started to focus its efforts at the level of policy-making and coordination. It has also started to promote the exchange of information among CILSS member states.

The Caribbean and the Pacific present similar models, but whereas one has long been successfully established, the other is still embryonic. Several ingredients may account for the different outcomes in the two regions. First, the relatively compact nature of the Caribbean, combined with its physical proximity to at least one of its major markets, eases communication flows and reduces transport costs, giving the region a comparative advantage in both traditional and high-value export crops. This has given governments in the region a strong commitment to research as a means of generating wealth and contributing to national development. It has also lowered the costs of collaboration between countries.

Second, and partly associated with its export tradition, the Caribbean, despite the small size of almost all its national research systems, has long possessed a number of relatively strong commodity institutes. Together, these have formed the building blocks for a strong regional research system. The Pacific, in contrast, possesses very weak national institutes. Indeed, as we have seen, some countries have no research capacity at all.

Third, both national and regional research in the Pacific is dominated by two "heavyweight" donors, each belonging, as do the region's countries themselves, to different linguistic and cultural traditions. In contrast, the sources of aid for CARICOM countries, while greater in number, are less schismatic, and there are fewer internal cultural and linguistic barriers in the Caribbean.

Many of the prerequisites for success must be effected at the policy level. Key among these is the political will to collaborate. Even with common ecological problems and zones, formal research collaboration is difficult if two countries are at odds. Creating the forum needed to coordinate research within a region depends on a political climate that encourages exchanges among countries, as was the case among the front-line states of Southern Africa. Where political ties remain stable over a long period, there can be sufficient trust to establish a regional research institute with pooled resources, as we have seen in the Caribbean.

BUILDING RESEARCH PARTNERSHIPS WITH REGIONAL AND INTERNATIONAL ORGANIZATIONS

The move towards regionalization of agricultural research depends upon three institutional bases. The first of these is the national agricultural research systems, which we regard as key. Second are the regional organizations that are growing in importance. Third, we have the international centers that are working increasingly in a partnership mode with national systems in the various regions. Successful regional research systems require greater planning, based on an understanding of the respective functions and comparative advantages of each of the national, regional, and international organizations involved.

There are two major functions to be assumed by regional research organizations. One is the coordination of research among a group of member countries or institutions, which is done by SACCAR, INSAH, and SPC. The second function involves defining a scope for research in the region and conducting that research on behalf of the region, as is done by CARDI, IRAZ, and IRETA. This second type of organization is probably unlikely to succeed unless the first type is also in place, but the reverse need not be true. There is some evidence to suggest that a coordinating body alone may be sufficient when there are favorable conditions for promoting the flow of technology both into and within a region. Either model (with or without the regional research institute) can be successful, but the additional presence of a regional institute that conducts research is more likely to cater to the needs of the smaller, more disadvantaged national systems.

International centers will need to work increasingly on a demand-driven basis as partners to NARS in regional initiatives. One way for them to do this is through the multicountry programs that work under the umbrella of regional organizations. Another is for them to act as initiators of regional task forces or collaborative regional research teams. A third is for them to pay greater attention to the liaison functions that researchers at both the centers and the NARS must perform.

For their part, small-country research systems need to consider and plan regional activities and collaborative research with international centers as a core part of their research programs. If they plan national research activities independently from the work of others in the region (be they other NARS, regional research organizations, or international centers) they can end up facing one of two possible outcomes: useful technology and information from outside will not "spill in" and contribute to economies of scope, and external partners who are bigger and have more resources may overwhelm a country's own national objectives and programs.

A clearly articulated national research policy, along with well-defined programs and priorities, is an essential tool for effective partnerships between small-country NARS and their regional and international partners. This applies equally to NARS in all countries. However, given the need for small-country NARS to function in partnership with larger institutions outside their country, it becomes a central aspect of national research policy and management.

Notes

1. The topic of networks is covered in a separate chapter. We prefer to consider networks as mechanisms for exchange and for establishing a division of labor among research organizations. This chapter is specifically concerned with *organizations* and *organizational frameworks* used at the regional level.

2. FAO currently executes 110 regional and interregional projects in agriculture, forestry, and fisheries. Its Near East Regional Commission on Agriculture has established six cooperative research networks on field food crops, sheep and goats, palms and dates, industrial crops, water use, and information and documentation. It also cosponsors the Association of Agricultural Research Institutions in the Near East and North Africa (AARINENA) and the Asian and Pacific Association of Agricultural Research Institutions (APAARI) (Beye 1992).

3. IICA has been instrumental in creating a number of regional and subregional networks, which have been increasingly vested with organizational responsibilities. Examples are PROCISUR, PROCIANDINO, PROCACAO, PROCICENTRAL, and PROCICARIBE.

4. The Caribbean Community and Common Market (CARICOM) has a membership of 12 countries, namely Antigua and Barbuda, Barbados, Belize, Dominica, Grenada, Guyana, Jamaica, Montserrat, St. Kitts and Nevis, St. Lucia, St. Vincent and Grenadines, and Trinidad and Tobago.

5. The member countries are Angola, Botswana, Lesotho, Malawi, Mozambique, Namibia, Swaziland, Tanzania, Zambia, and Zimbabwe.

6. The discussion that follows is derived from a regional analysis of small-country NARS in West Africa (see Gilbert, Matlon, and Eyzaguirre 1993).

7. According to Pollard (1988), the problems of the small South Pacific Island Nations are
   - few economies of scale
   - high degree of vulnerability to natural disasters and external shocks
   - narrow resource base
   - distant markets and high transport costs
   - poor market access for both exports and capital
   - limited independence in policy-making
   - limited skilled manpower

8. There are significant differences in the relations between the two institutes and their university hosts. IRETA relies on university staff for its research and is closely integrated with the School of Agriculture of the University of the South Pacific. CARDI retains a more independent posture from the Faculty of Agriculture, sharing few staff and facilities. This is in some part due to the separate historical development of the two institutions sharing a campus (see Parasram 1992; Wilson and Singh 1987).

10 Agricultural Research Networks in Small Countries

THE IMPORTANCE OF NETWORKS

The network, as a way of organizing the production of technology, information, goods, and services, is becoming a dominant feature of our times. In the days of Henry Ford, most of the parts of a motor car were produced and assembled in a single factory. Today, a network links factories around the world, each producing specific components that are assembled in another place. The trend towards networking is also reflected in the increasing decentralization, diversification, and specialization of research and development.

> In recent years, networking has come to be regarded as indispensable to the efficient conduct of scientific research, whether national or international and regardless of the level of economic development of the country or countries involved. In no field are research networks more important, or offer greater potential for increasing research effectiveness, than in applied agricultural research. This is particularly true in less developed countries where research networks can contribute greatly both to breaking isolation among scientists and, through sharing information and research tasks, to a more efficient use of scarce resources. (Hawtin 1991)

In research, networks are organizational mechanisms for linking scientists and institutions that are committed to sharing information or working together to solve common problems (Faris 1991). For all developing countries, whatever their size, networking in agricultural research offers a number of advantages:

- They promote efficiency through the pooling or sharing of scarce resources.
- They create common pools of knowledge that can be used to avoid duplication and repetition.
- They encourage technology and information spillovers from one country to another.
- They improve the quality of science by linking isolated national scientists to the global research community.
- They are necessary to address research issues that cross national boundaries.
- They facilitate training and the diffusion of new research technologies.

In theory at least, these benefits are even more useful for small countries than for large ones. For small countries, networking is a major way to organize and execute research. With their greater dependence on information and technology transfer as key functions of the national research system, small countries stand to benefit especially from the improved access to relevant technology and information provided by networks. Networks offer the small national research system a chance to tap expertise in areas where national capacity is either not feasible or not justified. They may also facilitate access to many useful services at the regional level that the small country cannot afford, including laboratory analysis and publishing outlets. Networks are particularly appealing to small countries because they offer a means of overcoming the problems of isolation to which scientists in such countries are especially prone. In many small-country NARS, the opportunity to participate in networks is a major factor keeping scientists in the system.

The benefits of networking are obvious; in practice, however, these benefits are difficult to secure. For the small national research system, networking can create special problems in research management. This chapter outlines some of these problems and examines the steps that managers can take to solve them.

TYPES OF AGRICULTURAL RESEARCH NETWORKS

Several classifications of research networks have been developed, generally focusing on the function that is performed. They range from simple information exchange to the allocation of tasks among member institutions.[1] The Special Program on African Agricultural Research (SPAAR) has identified the following categories of networks:

- *Information Exchange Networks.* These organize and facilitate the exchange of ideas, methods, and research results among participants.
- *Scientific Consultation Networks.* Participants in these networks focus research on common priority themes but conduct it independently.
- *Collaborative Research Networks.* Participants are involved in joint planning and monitoring of research on problems of mutual concern, and they share tasks.

All types of research networks share some key features. First, they promote an *exchange* between scientists or institutions. Second, the commodity most commonly exchanged is *information*. Third, collaborative research activities depend on an efficient exchange of information *on objectives and results.* Fourth, network effectiveness is largely determined by the *participants' ability to contribute* to, *and benefit* from, the information that is assembled and generated. Finally, *networks tend to evolve* from the basic and always essential function of information exchange to include coordination, resource exchange, and allocation. In the collaborative network aimed at technology generation, members assign each other the responsibility for tasks in the coordinated production of new knowledge and technologies to be shared among the membership. This is a more complex process to manage than other forms of networking.

NETWORK EVOLUTION AND INSTITUTIONALIZATION

In their initial stages, networks rarely have the means to provide the services and fund the exchanges expected by their members. In almost every case, other, well-established organizations create, host, and fund the network and provide the necessary forum to plan activities. Donor and technical-assistance agencies have often played this role. An outstanding case is Canada's International Development Research Center (IDRC), which has supported over 100 agricultural research networks.

Research networks can evolve to encompass a broader range of members and functions, and can develop new ways to allocate responsibilities among members. An important issue is the degree to which networks are institutionalized. They may start by linking individuals informally. Then, as exchanges and contacts become more regular, an institute may become a key player. Finally, the network may take permanent institutional form (Figure 10.1). Fully institutionalized networks tend to take on some of the characteristics of an organization.

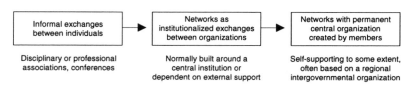

Figure 10.1. Evolution of networks

The Conference des Responsables de la Recherche Agronomique Africaine is an example where an annual meeting of francophone research leaders in Africa has evolved into a network of institutions and countries with policy support at the ministerial level. Initially, there were no permanent headquarters or staff to handle the affairs of the network; the presidency and headquarters rotated among the participating research leaders. Now, they have common activities, formal criteria for national membership, and a central secretariat with resources to manage network activities.

The trend toward institutionalization is not always positive. Donors place resources in a network, which research institutions then join in order to gain access to the resources. If access to these resources is the only motivation for institutions to participate, the network will disappear when the funds are exhausted. Network sustainability over the longer term depends on participants sharing common objectives and on their need to work together or to exchange information and expertise.

NETWORKS FROM A NATIONAL PERSPECTIVE

From a NARS perspective, there are three different types of network structure. The advantages and disadvantages of each are summarized in Box 10.1.

The first type can be described as *central-source networks*. Networks devoted to research on global staples and on many aspects of natural resource management are frequently hosted and managed by international agricultural research centers. These centers provide leadership, scientific capacity, technology, administrative support, and access to facilities and services. Examples are the International Maize Improvement

---

Box 10.1. Advantages and disadvantages of different network models

Networks managed by the IARCs are typical central-source networks, in which the technology-generating institution uses the network as a mechanism for carrying out its programs and transferring the technologies it generates. Provided the technology and resources available through the network are seen to be useful, enthusiasm about participating is often high. However, central-source networks do not always emphasize capacity building in the national system.

Building networks from an international center outwards, with the national system at the periphery, has several drawbacks. One is that external resources have to be provided continuously to maintain the network, allowing external donors to exert considerable influence on the direction of networked research. Another is that such networks promote exchanges between national focal points and the center but not among national research organizations themselves. Central-source networks might prove more sustainable once donor funding is withdrawn if they were to invest more resources in encouraging such direct exchanges among members. A third, and still more serious, drawback is that participation in central-source networks is extremely demanding on the limited resources of a small national system. Last, because larger countries tend to dominate these networks, small countries may find that network products and services are not relevant to their own national priorities.

In contrast, one advantage of regional networks is that they provide a forum at which policymakers and national research leaders can establish a regional policy for research and can collectively allocate responsibilities between national and regional institutions. In theory at least, such a forum places the small country on an equal footing with larger ones when setting research priorities, which is an important consideration for a country importing much of its technology. Collaboration in the Caribbean, for example, is facilitated by the fact that there is a regional policy framework on agriculture, food, and nutrition. This framework is used to establish the division of labor between regional and national institutions and to develop common research programs across them.

Professional networks are an effective and low-cost way for scientists to gain access to the latest knowledge, scientific methods, and technological developments in a given field. Because the key links are individuals and not institutions and because they function most often on a disciplinary basis, these networks do not receive as much attention from national research managers and donors as perhaps they should. Their key problem is that they are seen primarily as benefiting the individual scientist but not the institution. If these networks are to be used effectively by national research systems, an effort has to be made to channel the information received by participating individuals to the institution as a whole.

Network, based at the Centro Internacional de Mejoramiento de Maíz y Trigo (CIMMYT) in Mexico, and the African Trypanotolerant Livestock Network, based at the International Livestock Centre for Africa (ILCA) in Ethiopia.

Regional networks, based on intergovernmental regional organizations such as the Caribbean Community, the South Pacific Commission, or the Southern African Development Community, have forums where policymakers and NARS leaders can establish a regional policy for research and collectively allocate responsibilities between national and regional research. These networks operate under a common policy framework and have mandates that are derived directly from the priorities of member countries. To operate effectively, these networks must ensure that national priorities take research conducted elsewhere in the region into account.

The third category is the *professional network* or scientific society, established to promote a particular field of inquiry. Here, scientists themselves establish a network to foster the exchange of knowledge and to set scientific standards. The International Society of Sugar Cane Technologists (ISSCT), for example, draws members from the research and development sides of the sugar industry in all the major sugar producing countries. The society's specialist sections, publications, and regular meetings make it perhaps the major vehicle for the exchange and transfer of information and technology on sugarcane. This type of network, which unites scientists from developed and developing countries, is the oldest of the three and is the traditional means by which scientists exchange information.

Professional networks are a particularly effective, low-cost way for scientists and researchers to exchange information. Because the key links are individuals, not institutions, and because they function most often on a disciplinary basis, these networks do not receive as much attention from NARS managers and the donors who support research networks.[2] A key problem they pose for research managers and managers of scientific information, is that professional networks and societies are seen primarily as benefiting the individual scientist but not the institution he or she works for. If these networks are to be used effectively by NARS, an effort has to be made to channel the information received by participating individuals to the research programs and multidisciplinary work of the institution.

NETWORK MANAGEMENT PROBLEMS FOR SMALL-COUNTRY AGRICULTURAL RESEARCH SYSTEMS

Small countries are sometimes the "junior partners" in networks, but there are exceptions. Some small countries in the ISNAR study had built critical mass in a given commodity or research topic to such a degree that they were able to play a major role in a network. For example, Trinidad and Tobago plays an important role in global cocoa research. Through the work of its renowned sugar research institute, Mauritius plays a similar part in global research on this traditional export commodity. And the private and parastatal sectors in Honduras have made that country a focal point for global research on bananas. CARDI is a leader in the area of edible aroids (*Xanthosoma* and *Colocasia*). Sierra Leone has played a lead role in African rice research networks, and its work on cassava as a fresh vegetable and on sweet potato tubers and leaves is increasingly important to the wider region.

These success stories raise a question as to whether small developing countries can afford to play this leading role in research on selected commodities. We believe the answer is an unequivocal *yes*. Far from sapping strength from the rest of the system, the skills and scientific experience acquired in a single commodity can spill over to other areas. This is the case with the Mauritius Sugar Industry Research Institute and the Fundación Hondureña de Investigación Agrícola, both of which have been able to diversify their research interests to address a broader range of problems and domains than those implied by their mandate commodity (see also Chapter 6).

In most cases however, small countries have preferred to participate in networks rather than to lead them. And while networks are considered to be essential support mechanisms for small countries, the small size of the national research system makes it very difficult for them to participate and to manage their contributions. The irony is that those NARS that can most benefit from networks have the greatest difficulty in using them effectively.

The management problems and costs of networking are even greater when the research priorities of the network do not coincide with national priorities. Some networks are concerned mainly with disseminating the information and technologies that they produce to a group of client NARS. When the national system has few scientists who can spend a disproportionate share of their time at network meetings and executing network trials, participation in such networks has high opportunity costs.

The most obvious costs are in researcher and research management time: visits to network sites, participating in program management meetings and network steering committees, and receiving visitors, not to mention the actual research time spent in conducting the studies and experiments that are the focus of the network. Other less visible costs are in equipment, infrastructure, and the land that may be used in networked trials. There is no doubt about the benefits of networked activities, but are these benefits greater than what the system could have achieved with a well-focused national program? Unfortunately, at the national level, these costs and benefits have yet to be calculated, although there is a flood of documentation and evaluation reports, produced by network managers and sponsors on the networks themselves. This is an area of research management that will require closer examination in the future.

There is a simple first step that can help improve the management of networked activities. This involves looking at the number of networks to which the program staff within the NARS are formally committed. An example of this is in the Direction de la Recherche Agronomique (ARD) in Benin, which showed nearly a third of the national system's scientists involved in networks: a stark illustration of the overload problem (Figure 10.2).

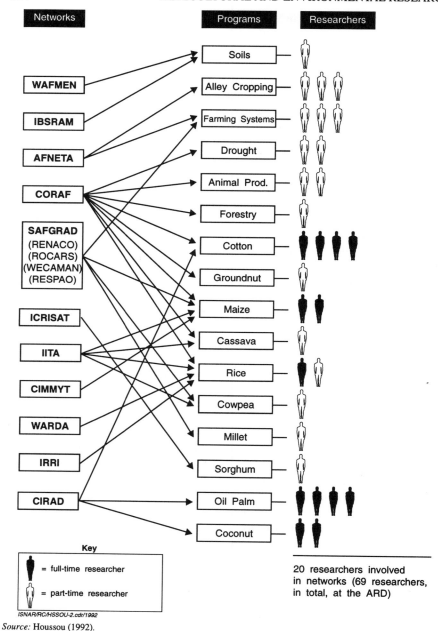

Figure 10.2. Benin's participation in networking

Source: Houssou (1992).

COPING WITH NETWORK OVERLOAD

There are now hundreds of agricultural research networks in the world, covering nearly all areas of research. In some areas, such as maize, there are several networks. To deal with potential involvement in this multitude of networks, the research manager must decide if a network can contribute to the quality and relevance of national research and whether participation would be a good investment of local resources. The criteria to be used in this decision include the following:

- the effect of participation on the priorities and balance of the overall national research effort;
- the focus of the network in relation to the core programs of national research organizations;
- the capacity of the national organization to participate in the network;
- the quality and quantity of information and materials exchanged within the network;
- whether these exchanges could benefit national research and development and how.

Attention to the type of network, levels of participation, and flows of information is also needed.

THE NETWORK TYPE

As we have seen, the type of network associated with different lead institutions has implications for the demands placed on the system as well as the probability that national research priorities will coincide with network priorities. Central-source networks, for example, encourage intensive participation by as many researchers and institutions as possible. For systems operating below a certain scale, with only a few researchers assigned to key topics and commodities, this can become a management problem, incurring heavy overheads and pitting the interests of the network against the needs of the institution to implement its own core programs. On the other hand, national priorities can be more effectively communicated to regional networks, and several regional organizations (such as SACCAR, CARDI, and IGADD) are actually based in small countries.

Small countries can manage more balanced participation in regional networks, because they allow greater inputs from national systems when priorities are set and responsibilities are assigned. ISNAR's study of small countries indicates that regional organizations play an important role by providing a forum where national policies and plans can be articulated to produce a regional research strategy.

In small countries, the small size of organizations and research programs means that much of the exchange of information happens through individuals as opposed to institutions. This means that professional networks, with their emphasis on individual exchanges, may be intrinsically more suitable for the small country. The function of the researcher in a small country is often that of a science and technology advisor as well as a developer of new technologies. He or she must often cover several commodities at once and can ill afford to go to every commodity-network meeting that falls within the scope of his or her program. The professional network may provide researchers with a short cut, at least allowing them to keep abreast of developments in methodology.

DECIDING ON THE LEVEL OF PARTICIPATION

Having decided on whether a network contributes to national priorities and objectives, research leaders need to consider at what level their system should participate in a network (Figure 10.3). Levels of participation range from collaborating intensively in research under a joint policy and program, to being a passive receiver of information, maintaining a *watching brief*, so to speak. A watching brief involves less commitment to a network and its activities; instead, the research organization observes and stays abreast of developments in the various networks and chooses carefully where it will commit its own resources.

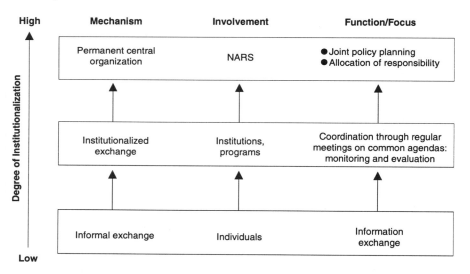

Figure 10.3. Network types and levels of participation

For those central-source networks devoted particularly to commodities that a country may produce in significant quantities but which are not the priority focus of national research, a small national system's best interests lie in keeping a watching brief with an eye open for training opportunities. This way the system can receive technology and information with little investment of its own resources. This may not please the coordinators of such networks, who often measure the success of the network by the degree of a country's participation, but from a national standpoint, intense participation may be more costly than it is worth. It can distort national priorities and make it difficult to implement national programs.

MANAGING THE FLOW OF NETWORK INFORMATION

Networks increase the flow of information to many small countries where the science and technology sector is small and underfunded. Given the importance of the information function in small national systems, this is a significant benefit. However, when the nodes of the network are individual researchers, there is a risk that the information they receive will not be institutionalized. This creates greater challenges and demands on information managers: working with scientists, they need to screen what is relevant, store what will be of more general use in the long term, and complement the network flows with other sources. In this case, networks also present a fine opportunity to link the management of information to the management of research programs (Eyzaguirre 1993b).

TRUST AND DIVISION OF LABOR

Participation in commodity networks that establish a division of labor among countries may pose problems for the small research systems, which must maintain more flexible research strategies and rely on a diverse set of small-scale actors. A division of labor also implies that many small countries would have to forego their capacity for maintenance research across several areas in order to concentrate resources on a key area in which they are the leader of the network. This is a desirable objective but only when countries can reasonably trust other countries to do research on their behalf and to make research results available. Building trust depends on countries increasing the knowledge of each other's programs, priorities, and capacities. National research systems may have a bit further to go in consulting with and learning about each other before they can establish the trust to make their agricultural sectors dependent upon research done elsewhere.

CONCLUSIONS

Networks are mechanisms for exchange and collaboration that small countries can use to extend the scope of their national research or to build long-term national capacity. The effectiveness of networks depends on the ability of each participating institution to produce information, to participate in joint activities, and to establish its own policies and structures for managing its involvement. Small NARS need the benefits that networking provides, but they may be at a disadvantage in managing their participation and contributions. They face at least one major conflict: networks must be focused to be effective, but a small NARS cannot afford to have more than a few intensive exchanges. It must be very selective about its participation in networks. When network participation overloads the national research scope, it can diminish a country's research capacity.

Currently, many networks encourage intensive participation from as many researchers and institutions as possible. For systems operating below a certain scale, with only a few researchers assigned to particular programs, this type of networking can become a management problem. Such networking puts the needs of the individual researcher against the institution's needs to run and implement its own programs.

The overhead costs associated with participation in networks are particularly burdensome for a small national research system. Committing precious staff time and other scarce resources to networks therefore requires careful evaluation of their potential costs and benefits. All too frequently, research managers fail to consider networks as primarily a mechanism for exchange. Viewing networks as exchange mechanisms rather than as organizational structures would allow managers to evaluate what their research institutions are contributing and what they are receiving via the network exchange.

Small national systems have the right to determine a level of participation in networks appropriate to their own needs and management capacity. In areas where a regional organization is able to act as a buffer between donors and IARCs on the one hand and national research systems on the other, small countries may stand a better chance of having their voices heard when network research priorities are established. Central-source networks, in particular, should be encouraged to develop more flexible levels of national participation, ranging from intensive collaboration to maintaining a watching brief. There is also the growing expectation that national research systems will increasingly take over much of the planning and coordination of networks now managed by the international centers (Plucknett, Smith, and Ozgediz 1990: 175).

Research leaders and policymakers in small countries need to be selective in their network participation. This means that many network management decisions should be treated as part of the decision-making for national programs. If properly selected and managed, network participation can enable researchers to receive the information and professional rewards they deserve. It can also give national institutions the opportunity to acquire additional capacity and extend their scope to meet national goals and contribute to regional research.

Notes

1. The book *Networking in International Agricultural Research* by Plucknett, Smith, and Ozgediz (1990) contains a comprehensive review of the structure and function of agricul-

tural research networks. Valverde (1988) produced a review that considers issues in the establishment and management of networks. This chapter is more concerned with networks as a mechanism for exchange and with networking as a mechanism for either decentralizing or linking research units engaged in technology generation.

2. Anderson (1992) has noted that professional associations as research networks are often—and unjustifiably—ignored.

11 Conclusion: Lessons from Small Countries

BEYOND THE CONVENTIONAL WISDOM

In this book, we have focused on three objectives. One is to synthesize the findings of research leaders in small developing countries who, as part of our study, examined the conditions and strategic options for strengthening their national research systems. The second is to propose a new portfolio approach to research policy and organization that supports flexible and innovative organizations. The third is to use these findings to advocate greater support and recognition for national agricultural research in small developing countries.

There is no substitute for a national research capacity in a country. If a country has a national policy for developing its agricultural sector and for managing its natural resource base, then it will need some research capacity to identify and evaluate its options. In order to build that essential capacity, small countries have to do things differently. Our analysis of the conditions and options for building a viable research system has led us to reject the conventional wisdom on what national research systems are and what they should do (see Table 11.1).

Conventional wisdom has it that small countries have a smaller range of institutions involved in research. This proved not to be the case. We found that a diverse set of actors and, indeed, institutional diversity provided the base upon which the more robust small-country NARS are established.

Conventional wisdom has it that research efforts in small countries should be consolidated under a single organization. This also proved not to be the case. A major finding of this study was that broadening the institutional base and promoting greater flexibility in the structure of national agricultural research systems is the key to their success, or even to their survival in some cases.

Common sense tells us that research systems in small countries should narrow their scope to do less. However, our cases showed that this was not possible in reality. The answer is not to do less, but to do things differently. The key to handling the inevitably broad scope that national research on agriculture and natural resources entails is for small countries to emphasize a broader set of NARS functions and to distribute them across a range of institutions.

nonsense [handwritten]

Table 11.1. Beyond the Conventional Wisdom

The Conventional Wisdom on Small Countries	Findings from the Study
• Small countries have fewer organizations involved in research	• Small countries have diverse institutional actors that together comprise the NARS
• Disparate research activities should be consolidated in a single organization	• Consolidation into one organization is often not advisable—an effective scale of research is built on diversity
• Given limited resources, small-country NARS should narrow their scope to cover fewer areas	• Narrowing the scope of national research is difficult given policy demands and changes in the agricultural and natural resource sectors
• Small countries will be "technology borrowers" that do not require scientists with high-level training	• Intelligent use of external knowledge requires a sophisticated scientific research capacity that can relate local needs to available technology
• Small-country research systems carry out fewer and less complex functions	• The smaller the system, the more complex the functions it will perform; instead of doing fewer things, small-country NARS should do things differently—they need to emphasize a broader set of functions
• Research in small countries is either vulnerable or not viable because it cannot break its dependence on donors and external agencies	• All research is increasingly done in networked and interdependent modes; small countries can take a more active role in managing partnerships with donors and other external agencies

What really sustainable intellectual [handwritten]

When agricultural research is evaluated, the traditional focus is on technology generation, where small countries are involved mainly in testing technologies developed by international research centers. Our focus has been on technology *scanning,* which calls for access to a wide range of sources and information on technology.

Sustainability of a national agricultural research system is often defined as breaking its dependence on donors and external sources of technical assistance. What we have discovered is that the strategy small countries should pursue is not to break with donors and external technical assistance; instead, they need to strengthen policy interactions in order to focus and derive greater benefit from donor projects and the activities of international organizations.

What does this mean? [handwritten]

MANAGING NATIONAL RESEARCH AS A DIVERSIFIED INSTITUTIONAL PORTFOLIO

In the preceding chapters, we have presented examples and analyses of the policies, organizational strategies, and functions that embody a new approach: we consider research as a portfolio of institutionalized activities responding to demands from different domains. This portfolio approach can be used to mobilize a range of diverse institutional actors. It also provides a method for analyzing institutional comparative advantage, which makes it possible to evaluate when and where consolidation makes sense and to assign responsibility for different functions. In most cases, institutional diversity across domains provides better opportunities to gain access to a broader range of funding, technology, and information sources and to improve linkages with clients.

This book has emphasized the appropriateness of such an approach for small countries with fixed constraints on organizational growth and diseconomies of institutional development. But in the course of this study, it became increasingly obvious that this approach would be of general benefit to most developing countries. Here, as in the international business community, it is the small firms (or countries) that can teach the larger ones how to innovate and how to remain flexible to take advantage of new opportunities in an increasingly competitive and rapidly changing environment.

THE POLICY LEVEL: THE KEY TO MAKING INSTITUTIONAL DIVERSITY WORK

Who manages the national research portfolio? This is not a task that can be accomplished by a manager within one research organization; it must be done at the research policy level. On the other hand, policymakers alone cannot establish and manage the portfolio, either. It requires determining risks, returns, and levels of investment, which research leaders and policymakers need to do together.

The need for research to interact with policy is crucial. To make institutional diversity work, there has to be good coordination and accountability, which, along with policy-making, priority setting, and monitoring and evaluation, require good links to policy. National research coordinating councils or national directorates for research and development are useful mechanisms for this function if the right participation can be assured, but many small countries have no mechanism or forum where this research policy function can take place. While other private and parastatal institutions within the research system are a welcome addition to the national capacity, they cannot take the place of public-sector units in performing these policy and coordination functions.

In many cases, research institutions are not well connected to the policy and decision-making levels within the agricultural and natural resource sectors. Some of the government research services may be too small to generate technologies and are considered marginal to the process of managing technology. In such cases, research departments need to be placed at higher levels within the public structures for science and technology. If small-country NARS are to have an impact despite the small size of their research units, they must work more closely with policymakers to mobilize and coordinate a range of institutional actors.

DEMANDS FOR BROADER SCOPE CAN BE MET THROUGH BROADER FUNCTIONS

NARS in small countries will have to define their functions in different ways than larger systems. Making technology available to improve agricultural productivity and efficiency is clearly a central responsibility of research, but it is not necessarily dependent upon experimentation. NARS in small countries may be less intensive in commodity research and technology generation, but more sophisticated in the scanning and assessment of world science. In order to match technologies to the problems and policies of their countries, NARS may need to devote greater effort to studying the natural resource base and existing production systems. The complexity of technology scanning, transfer, and adaptation has been grossly underestimated by many donor agencies, who urge research systems in smaller countries to be "technology borrowers". We have emphasized the need for NARS to develop and maintain a sophisticated scientific capacity in order to be intelligent borrowers. This means greater investment in human capital.

As we considered the full range of NARS functions, from the traditional one of experimentation through to policy advice, coordination, management of linkages and information, and regulation, it became apparent that in small countries these functions are closely interrelated. Indeed it confirmed our general premise that as the scale of the institutions becomes smaller, the complexity of the functions they are expected to perform actually increases. Our studies showed that researchers are inevitably expected to perform many functions beyond running experiments and studies, but this holds many potential rewards for researchers and research managers as they assume more responsibility for the entire technology development process.

Among the broad functions identified in the portfolio approach, there are three that lie at the core and that are interrelated. They are

- technology adaptation, screening, and testing
- policy advice and coordination
- information analysis and synthesis

In small countries, the performance of any one function requires the ability to do the others. Even technology screening and testing requires a broader natural resource management and systems perspective. Screening available technologies and acquiring a thorough knowledge of the country's natural resource base is a prerequisite for adaptation. You need to know what you want, and to be able to communicate it to outside agencies.

The findings of ISNAR's study have indicated key ways that NARS in small countries can participate as full partners in the global effort to increase the productivity of agriculture and safeguard the natural resource base for the future. Our studies have shown that the management capacity needed to capture spill-ins, to use networking, and to work in multicountry partnerships is great indeed. We have emphasized the need to be sophisticated and innovative in organizational strategies as well. Small countries are well placed to be innovators in this regard, for they have little choice.

LEADERSHIP

The functions that are crucial to research systems in small countries are those that establish a capacity for decision making and technology management at the heart of the national system. This creates opportunities for national and scientific leadership; small size reduces distance. In a country like Guinea-Bissau, for example, it is possible to assemble all the relevant policymakers and research leaders in one room to arrive at a consensus, as was done in 1989 (ISNAR 1989b). Researchers in small countries have the opportunity to take a holistic approach to agricultural technology systems that only the most senior research managers and policymakers might have in a large country (see Box 11.1).

LESSONS FOR NATIONAL AGRICULTURAL RESEARCH SYSTEMS

Managing human, financial, and physical resources is perhaps not so different between small and large countries. The differences are related more to types of organizations than to the size of countries. For this reason, our lessons are aimed mainly at the level of research policy.

The strategies and functions described in the preceding chapters illustrate a range of opportunities for meeting the increasing demands placed on research. We have shown that emerging issues, such as diversification, new high-value crops, and new techniques (including biotechnology), cannot be ignored, even in a small country with few resources. Indeed, growing concerns with natural resource management and conservation require every country to have a national research capacity in place. Increasing the stock of knowledge on the country's national resources expands the range of options available for using them more efficiently and productively. Decisions about the best use of a country's resources are national decisions, but to make the best decisions, policymakers need the best information, which should come from their national

---

Box 11.1. Leadership in a Small-Country NARS: Focusing on a Holistic Approach to Technology Development and Accountability

One research director of a small country with a distinguished research tradition described in the following terms what his work means to him.

If I had made my career in a country with a large institute and plentiful resources, I probably would have become a leading world authority in one particular aspect of the crop on which I have devoted most of my life's work. Because I have devoted my career to increasing the productivity of the farmers in my own small country I have had to adopt a broader approach to research. I am accountable to farmers not only for doing research but for the final result: that they have technologies and information that allow them to be more productive and profitable.

I am responsible for the whole technological development process. It begins by keeping abreast of the latest developments in science and technology that may be relevant to our national crop. For this, a researcher must have scientific credibility in order to liaise with the international science community. At the same time, I must understand the needs of our farmers and our national policy. When there is a new scientific or technological development that can be adapted to our conditions and meets our needs, I play an active role in all aspects of the process from research to technology to adoption. I can derive satisfaction from seeing my initial ideas and scientific understanding develop into practical technologies reflected in increased production for our farmers. This holistic approach from the scientific community to research and development to production, bringing together the scientist, policymaker, and producer, can bring many rewards to the development-oriented scientist. In a small country, there are many opportunities for a scientist to see his or her ideas put into practice and to gauge their impact.

Source: Unpublished rapporteur notes on the closing discussions of the International Workshop on Agricultural Research in Small Countries, Réduit, Mauritius, 1992.

---

research system. Small-country research systems must be outward looking in their use and application of technology and information, but they need to strengthen their national capacity to define their own priorities and establish a clear national policy for research.

Research should be used for policy-making as well as for technology development. Countries can be more effective users of available knowledge and technology if their understanding of existing production systems and constraints is improved. At the same time, they can also make contributions to global scientific knowledge by applying and contributing the experience and understanding that arises from their own situation. To do this, they need to ensure that locally produced research is adequately documented and applied. That this is often not the case is one widespread problem in small-country research systems that urgently needs to be addressed.

LESSONS FOR DONORS AND EXTERNAL PARTNERS

Partnerships with external agencies, such as regional research organizations, international research centers, neighboring national systems, and intergovernmental organizations are crucial if small countries are to acquire the essential research capacity they need. However, small countries that do not have clear policies and strategies of their own can easily be overwhelmed or bypassed by larger research institutions. We have seen how this is all too often the case in research networks and regional research collaboration. What we propose is to treat the linkages with external research partners as a core part of the national research policy and planning process, subject to the same decisions as national investments. One lesson from ISNAR's study of small countries that is relevant to all, is that effective collaboration begins with the formulation and management of research policies and plans at the national level.

Donor and technical agencies are likely to remain an essential component of the resource base for research in small countries, and many of these are rightly enthusiastic about moves towards greater intercountry collaboration, networking, and regional approaches. There are indisputable benefits and economies that can be made by sharing and pooling resources in regional initiatives and networking, but it must be remembered that this requires additional investments in management and institutional capacity.

Donor agencies need to consider how research management functions can be strengthened at the national level in order to establish effective research partnerships among countries. International centers and regional organizations working with small countries need to pay greater attention to national research policies. Where countries have not developed them, an initial investment in regional collaboration must include assisting countries in formulating national research policies and priorities.

This study has emphasized that smaller systems are not necessarily less complex. On the contrary, the functions of a smaller system are often more complex. This means that research management functions are more crucial and widespread at all levels of the system and it is often difficult to delegate them to a higher level. In a small NARS, the importance of management within the overall system cannot be overstated, and an emphasis on the importance of management will become increasingly evident in all developing countries.

Combining management functions with a broader and more sophisticated capacity in scientific scanning, information management, and advising science policy requires a higher level of training than what many donor agencies have hitherto considered. Smaller research systems cannot function on only short-term and lower levels of training. Because many of the research programs in small countries consist of only one to five persons, donors should consider an increase or change in their support to invest in building a broader capacity for analytical science. This means a shift in resources away from short-term, narrow technical training to broader, more basic scientific training for some scientists. A group of scientific decision makers who could understand agricultural options would provide the policy core that would make small countries more effective partners in the global research system.

THE NEW RESEARCH ENVIRONMENT: OPPORTUNITIES FOR SMALL COUNTRIES AND SMALL SYSTEMS

The concluding chapters of this book have discussed the implications of regional initiatives and networking to small countries. The key for these countries to be in a position to take advantage of the new opportunities provided by this globalization is for them to use the new modes of research as part of their national plan and as a means of implementing national goals. The focus on regional collaboration only heightens the need for effective, flexible research capacity at the national level. Like the more successful business firms of the nineties, small-country NARS will have to use innovative approaches to policy and management that parallel the global trends in research and development within both the private and public sectors (see Box 11.2).

Box 11.2. Small Organizations Innovate by Mobilizing Others

A manager of a leading, very successful small company in Germany describes his strategy as follows (Peters 1992):

> We don't want to be big, we want to be innovative, to focus on know-how, not putting up more bricks, more buildings, big factories, etc. We want to emphasize the innovative power of the organization. Being small means "doing in-house what we can do better than others and giving to outsiders anything else". Relying on others to make part of the product or generate the information means your small organization must manage those networks by maintaining standards, and treating outside (sub-contracted) work like production of your own. (pp 298–299)

> Most of yesterday's highly integrated giants are working overtime at splitting into more manageable, more energetic units. They are building up new productive and research capacity not by acquiring new facilities and companies but via alliances with all sorts of partners of varying sizes and shapes. In the newest and most forward looking of industries in information technology and biotechnology, new forms of networked organizations are emerging that can deal with rapid technological and scientific progress, and new producers and markets. (p 303)

In many ways small countries are especially well suited to facing the challenges of the future. In many systems, researchers already act in multiple roles, as we have shown in Box 11.1. Because the system is so small, contacts with the policy-making levels in government are easier to make and consensus is easier to reach. While larger systems are downsizing, small systems are already lean and already dealing with innovative ways of coping with an uncertain future.

In the past, national agricultural research has concentrated on producing technologies for the agricultural sector, but it can no longer afford to do that. Instead, national systems will have to increasingly weigh the level and direction of their agricultural research, and they will have to balance the different demands of borrowing as well as producing agricultural technologies. For small countries, the way they address these sometimes conflicting demands on their systems may determine whether they survive at all.

The history of research in the developing world owes much to the seminal role played by research institutions in small countries like Sierra Leone, Trinidad and Tobago, Mauritius, Fiji, and others. As research becomes an increasingly global enterprise, new and greater efforts will be needed to ensure that research systems in small countries do not become the weak link in the chain. This book draws upon their experience to provide a strategy to strengthen those global links.

Annexes

Annex 1. Explanation of the Type of Institutions Executing or Supporting Research

Type of Institution	Definition
Agricultural Research Councils	National research coordinating and planning entities based in the public sector: NARCC, Sierra Leone
Research Foundations	Organization mandated by government to operate as a private agency: Jamaica Agricultural Development Foundation, Fundación Hondureña de Investigación Agrícola
French Tropical Research Institutes	This refers to the CIRAD and ORSTOM complex of institutes
Ministries	Ministries of agriculture, natural resources, or primary industries where no specific research units could be identified, such as in Belize, Dominica, St. Lucia, Trinidad and Tobago, and Mongolia
Government Agricultural Research Institutions	Normally research institutes, departments, or divisions within ministries of agriculture, natural resources, or primary industries
Government Research Institutes	Institutes with scientific mandates beyond research on crops and livestock. Examples: the Institut Superieur d'Etude et de Recherche Scientifique et Technologique (Djibouti), Centre National de Recherche Océanographique et des Pêches (Mauritania), Direction de Nutrition et de Technologie Alimentaire (Togo), Wau Ecology Institute (Papua New Guinea), and the Institute for Applied Science and Technology (Guyana)
Multinational Agribusiness	Multinational agribusiness with defined research units. Examples: the Rubber Research Institute (Liberia), Chiquita Brands Co. (Honduras), and Levers Solomons Ltd. (Solomon Islands)
National Agribusiness	National agribusiness with formal research activities. Example: Caroni Ltd. in Trinidad and Tobago
Parastatal Organizations	Parastatal organizations with research units, normally attached to commodity boards. Examples: the Sugar Research Institute (Jamaica), Coffee Research Institute (El Salvador), Papua New Guinea Oil Palm Research Association
Nongovernment Organizations	Nongovernmental organizations chartered in a country. Example: the Rural Industries Innovation Centre (Botswana)
Regional Organizations	Regional organizations with mandates beyond agricultural research. Examples: Inter-American Institute for Cooperation on Agriculture, CARICOM, Southern African Development Community (SADC), and SACCAR
Regional Research Organizations	Institutions mandated to carry out agricultural research at the regional level. Examples: IRAZ, CARDI, IRETA. Each is counted only once but they work in several countries
Regional Universities	Regional universities, such as the University of West Indies, the University of the South Pacific, and the Pan American School work in a number of countries but have been counted only once
National Universities	Universities such as the University of Guyana, Njala University of Sierra Leone, PNG University of Technology which have mandates and programs that define them as core institutions of the NARS

Annex 2. Checklist of Agricultural Research Organizations in Small Countries

Country	Primary/Secondary*	Research Organizations	Institute Type**
Barbados	Primary	1. Crops Research, Development and Extension	GRA
		2. Livestock Research, Development and Extension Division (MOA)	
		3. Caribbean Agro-Industies Ltd.	MNAB
		4. Sugar Technology Research Unit (Barbados Sugar Industry Ltd.)	NAB
		5. Agronomy Research Unit (Barbados Sugar Industry Ltd.)	
		6. Barbados Sugar Cane Variety Testing Station	PARA
		7. West Indies Sugar Cane Breeding Station	RRO
	Secondary	8. Soil Conservation Division (MAFF)	GRA
		9. Caribbean Agricultural Research and Development Institute	RRO
		10. Biology Department, University of the West Indies	RUNI
Belize	Primary	1. Research Division of the Department of Agriculture	GRA
	Secondary	2. Caricom Farms Ltd.	PARA
		3. Humming Bird Hershey Ltd.	MNAB
		4. Belize Sugar Industry Limited	PARA
		5. Belize Agri-Business Company	
		6. Caribbean Agricultural Research and Development Institute	RRO
Benin	Primary	1. Comité National de la Recherche Agronomique	ARC
		2. Direction de la Recherche Agronomique	GRA
		3. Unite de Recherche Zootechnique et Vétérinaire	
	Secondary	4. Le Conseil National de la Recherche Scientifique et Technique	ARC
		5. Centre Béninois de la Recherche Scientifique et Technique	GRI
		6. Faculté des Sciences Agronomiques	UNI
Bhutan	Primary	1. Research Unit, Dept of Agriculture (MOA)	GRA
		2. Research & Training Division, Dept of Forestry (MOA)	
		3. Research Unit, Dept of Animal Health (MOA)	
	Secondary	4. Natural Resources Training Centre	
		5. Service Unit, Dept. of Agriculture (MOA)	
		6. Service Unit, Dept. of Forestry	
		7. Service Unit, Dept. Animal Husbandry (MOA)	
		8. Veterinary Research Centre	
		9. District Extension Units, Ministry of Home Affairs	
Botswana	Primary	1. Department of Agricultural Research	GRA
	Secondary	2. Rural Industries Innovation Centre	NGO
		3. Botswana Livestock Development Corporation	PARA
		4. Botswana Development Corporation	
		5. University of Botswana, Faculty of Agriculture	UNI
Burundi	Primary	1. Institut des Sciences Agronomiques du Burundi	GRA

Annex 2. (continued)

Burundi *continued*	Secondary	2. Institut de Recherche Agronomique et Zootechnique	RRO
		3. University of Burundi	UNI
Cape Verde	Primary	1. Instituto Nacional de Investigação Agraria	GRA
		2. Centro de Desenvolvimento Pecuario	
		3. Instituto Nacional de Investigação Pescária	GRI
Central African Republic	Primary	1. System National de la Recherche Agronomique	GRA
Chad	Primary	1. Bureau de la Recherche Agromique	GRA
		2. Laboratoire de Recherches Vétérinaires et Zootechniques	
	Secondary	3. L'Institut de Recherche du Coton et des Textiles	FTRI
		4. Institut Universitaire des Techniques de l'Elevage	GRA
Comoros	Primary	1. Research & Development Unit, Federal Centre of Rural Development	GRA
Congo	Primary	1. Centre de Recherche Agronomique de Loudima	GRA
		2. Centre de Recherche Vétérinaire et Zootech	
		3. Directn. de la Recherche-Develop., Formation et Vulg	
		4. Centre National d'Etudes des Sols	
		5. Centre National de Pisciculture de Djoumouna	
		6. Centre National de Semences Ameliorées	
		7. Centre d'Etude sur les Ressources Végétales	GRI
	Secondary	8. Office de la Recherche Scientifique et Technique d'Outre-Mer (ORSTOM)	FTRI
		9. Centre Technique Forestier Tropical	
Djibouti	Primary	1. Institut Superieure d'Etude et Rech. Sc.et Techn.	GRI
Dominica	Primary	1. Ministry of Agriculture (Policy)	GRA
	Secondary	2. Windward Island Banana Growers' Association	NAB
		3. Caribbean Agricultural Research and Development Institute	RRO
El Salvador	Primary	1. Fundación Salvadoreña para el Desarollo	FOU
		2. Centro de Desarrollo Ganadero	GRA
		3. Centro de Desarrollo Pesquero	
		4. Centro de Recursos Naturales	
		5. Centro Nacional de Tecnología Agropecuaria	
		6. Instituto Salvadoreño de Investigaciones del Café	PARA
Equitorial Guinea		*NO RECOGNIZABLE RESEARCH INSTITUTIONS*	
Fiji	Primary	1. Research Division of the Department of Agriculture	GRA
		2. Research Section, Animal Health & Production Division	
		3. Silviculture Section (Ministry of Forestry)	
		4. Pine Section (Ministry of Forestry)	
		5. Timber Utilisation Section (Ministry of Forestry)	
		6. Sugar Cane Research Institute	PARA

Annex 2. (continued)

Fiji *continued*	Secondary	7. Fisheries Division (Ministry of Primary Industries)	GRA
		8. University of South Pacific	RUNI
Gambia	Primary	1. National Agricultural Research Board	ARC
		2. Department of Agricultural Research	GRA
		3. Research Unit (Department of Livestock Services)	
	Secondary	4. Research Unit (Department of Planning)	
		5. Research Unit (Department of Water Resources)	
		6. Research Unit (Department of Fisheries)	
		7. Research Unit (Department of Forestry)	
		8. Research Unit (Department of Wildlife Conservation)	
Grenada	Primary	1. Crop Protection Unit, Ministry of Agriculture	GRA
	Secondary	2. Technical Unit, Grenada Cocoa Board	NAB
		3. Nutmeg Association	
		4. Caribbean Agricultural Research and Development Institute	RRO
Guinea-Bissau	Primary	1. Departamento de Estudos e Pesquisa Agricola	GRA
Guyana	Primary	1. National Agricultural Research Institute	GRA
		2. University of Guyana	UNI
		3. Research Division, Guyana Sugar Corporation Ltd.	PARA
	Secondary	4. Livestock Development Company	
		5. National Dairy Foundation	
		6. Caribbean Agricultural Research and Development Institute	RRO
		7. Institute of Applied Science & Technology	GRI
Honduras	Primary	1. Fundación Hondureña de Investigación Agrícola	FOU
		2. Crop Research Department	GRA
		3. Livestock Research Department	
		4. Centro Nacional de Investigación Forestal Aplicada	
		5. Research Department, Standard Fruit Company	MNAB
		6. Research & Technical Department, Chiquita Brands Company Ltd.	
		7. Research Department, Instituto Hondureño del Café	PARA
	Secondary	8. Corporación Hondureña para El Desarrollo Forestal	
		9. Pan American School	RUNI
		10. Escuela Nacional de Ciencias Forestales	UNI
Jamaica	Primary	1. Scientific Research Council	ARC
		2. Jamaican Agricultural Development Foundation	FOU
		3. Department of Agricultural Research & Development	GRA
		4. Research Department, Jamaican Banana Board	PARA
		5. Research Unit, Jamaican Coconut Industry Board	
		6. Sugar Research Institute	
		7. Research and Extension Division, Cocoa Industry Board	

Annex 2. (continued)

Jamaica *continued*	Secondary	8. Jamaican Soil Survey Laboratory	GRA
		9. Citrus Growers Association	NAB
		10. Coffee Industry Development Co.	PARA
		11. Interamerican Institute for Cooperation on Agriculture	RRO
		12. Caribbean Agricultural Research and Development Institute	
		13. University of the West Indies	RUNI
		14. College of Agriculture	UNI
Kiribati	Primary	1. Agricultural Division (MOA)	GRA
		2. Atoll Research and Development Unit	
Laos	Primary	1. State Committee for Science & Technology	ARC
		2. Centre National de Recherche Agronomique	GRA
	Secondary	3. Department of Livestock and Veterinary Services	
		4. Department of Forestry and Environment	
Lesotho	Primary	1. Agricultural Research Division, Ministry of Agriculture, Cooperatives and Marketing	GRA
Liberia	Primary	1. Central Agricultural Research Institute	GRA
		2. Liberia Rubber Research Institute	PARA
	Secondary	3. University of Liberia	UNI
Mauritania	Primary	1. Centre National de l'Elevage et Recherche Vétérinaire	GRA
		2. Centre National Recherches Agronomiques & Développement Agricole	
		3. Laboratoire Central Vétérinaire	
		4. Centre National Recherches Océanographique & des Peches	GRI
	Secondary	5. Laboratoire National d'Analyse des Sols	GRA
		6. Laboratoire d'Etudes et de Recherches Geographiques	GRI
Mauritius	Primary	1. Department of Agriculture and Scientific Services	GRA
		2. Mauritius Sugar Industry Research Institute	PARA
	Secondary	3. School of Agriculture, University of Mauritius	UNI
Mongolia	Primary	1. Council of Agricultural Science	ARC
		2. Ministry of Agriculture and Food Industry	GRA
Namibia	Primary	1. Agricultural Research Division (MAWRD)	GRA
	Secondary	2. Directorate of Veterinary Services (MAWRD)	
		3. Directorate of Forestry (MAWRD)	
		4. National Institute for Social and Economic Research, University of Namibia	UNI
		5. Namibian Economic Policy Research Unit	GRI
		6. First National Development Corporation	PARA
		7. Desert Ecological Research Unit	GRI
Nicaragua	Primary	1. Instituto Nacional de Technología Agropecuaria	GRA
Panama	Primary	1. Instituto de Investigación Agropecuaria de Panamá	GRA
	Secondary	2. Faculty of Agronomy of the University of Panama	UNI

Annex 2. (continued)

Papua New Guinea	Primary	1. National Agricultural Council	ARC
		2. Agricultural Research Division (MOA & Livestock)	GRA
		3. Wau Ecology Institute	GRI
		4. PNG Oil Palm Research Association	PARA
		5. PNG Cocoa and Coconut Research Institute	
		6. PNG Coffee Research Institute	
		7. PNG Sugarcane Research Centre	
		8. University Technology	UNI
	Secondary	9. Department of Forests	GRA
		10. Fisheries Department	
Paraguay	Primary	1. Department of Agricultural, Livestock and Forestry Research & Extension	GRA
		2. Department of Agricultural Economic Research	
		3. Instituto Agronómico Nacional	
Rwanda	Primary	1. Institut des Sciences Agronomiques du Rwanda	GRA
	Secondary	2. Institut de Recherche Agronomique et Zootechnique	RRO
		3. National University of Rwanda	UNI
São Tomé e Principe	Primary	1. Centro de Culturas Alimentares de Mesquita	GRA
		2. Estacão Experimentale Agronomica de Poto	
Seychelles	Primary	1. Agricultural Promotion Division (MOA & Fisheries)	GRA
		2. Fisheries Division (MOA & Fisheries)	
Sierra Leone	Primary	1. National Agricultural Research Coordinating Council	ARC
		2. Rokupr Rice Research Station	GRA
		3. Institute of Agricultural Research, Njala	
	Secondary	4. Land and Water Development Division	
		5. Institute of Marine Biology and Oceanography	GRI
		6. Planning, Evaluation & Monitoring Services Division	
		7. University of Sierra Leone, Fourah Bay	UNI
		8. Njala University College	
Solomon Islands	Primary	1. Department of Research, Ministry of Agriculture and Lands	GRA
	Secondary	2. Forestry Division of the Ministry of Natural Resources	
		3. Levers Research (Yandina) of Levers Solomons Ltd.	MNAB
Somalia	Primary	1. National Agricultural Research Institute	ARC
		2. Livestock and Environment Research Institute	GRA
	Secondary	3. Serum and Vaccine Institute	
		4. Faculty of Agriculture, Mogadishu University	UNI
St. Lucia	Primary	1. Ministry of Agriculture	GRA
	Secondary	2. Windward Islands Banana Growers' Association	RO
		3. Caribbean Natural Resources Institute	RRO
		4. University of West Indies	RUNI

Annex 2. (continued)

St. Vincent	Primary	1. Research Unit, Department of Agriculture (MOA)	GRA
	Secondary	2. Windward Islands Banana Growers' Association	RO
		3. University of West Indies	RUNI
		4. Caribbean Agricultural Research and Development Institute	RRO
Suriname	Primary	1. Landbouwproefstation	GRA
	Secondary	2. Foundation for Mechanised Agriculture	FOU
		3. Practical Research on Rice	PARA
		4. Foundation for Experimental Farms	
		5. Foundation for Agricultural Development Plan Commewijne	
		6. SURLAND	
		7. Foundation for Experimental Gardens in Suriname	
		8. Victoria	
		9. University of Suriname	UNI
Swaziland	Primary	1. Agricultural Research Division (MOA & Cooperatives)	GRA
	Secondary	2. Veterinary Diagnostic Laboratory	
		3. SWAZICAN (Swaziland Fruit Cannery)	NAB
		4. Usuthu Pulp Company Limited	
		5. Royal Swaziland Sugar Company	
		6. Swaziland Sugar Association	
		7. Mhlume (Swaziland) Sugar Company Ltd.	
		8. Swaziland Cotton Board	PARA
		9. Faculty of Agriculture, University of Swaziland	UNI
		10. Social Science Research Unit, University of Swaziland	
Togo	Primary	1. Direction Nationale de la Recherche Agronomique	ARC
		2. Institut National des Cultures Vivrières (DRA)	GRA
		3. Direction de la Protection des Végétaux	
		4. Institut des Plantes à Tubercules	
		5. Institut National des Sols	
		6. Centre de Recherche et d'Elevage d'Avetonou (CREAT)	
		7. Direction de la Nutrition et de la Technologie Alimentaire	GRI
	Secondary	8. Office de Recherche Scientifique et Technique d'Outre-Mer (ORSTOM)	FTRI
		9. Institut de Recherche du Café et du Cacao (IRCC)	
		10. Institut de Recherche du Coton et des Textiles Exotique (IRCT)	
		11. Institut National de la Recherche Scientifique	GRI
		12. Ecole Superieure d'Agronomie	UNI
Tonga	Primary	1. Agricultural Research Division, Department of Agriculture	GRA
Trinidad and Tobago	Primary	1. Ministry of Food Production and Marine Exploitation	GRA
		2. Ministry of Environment and National Service	

Annex 2. (continued)

Trinidad and Tobago *continued*	Secondary	3. Sugar Cane Feeds Centre	GRA
		4. Institute of Marine Affairs	GRI
		5. Caroni Sugar Research	NAB
		6. Caribbean Industrial Research Institute	PARA
		7. Caribbean Agricultural Research and Development Institute	RRO
		8. University of the West Indies	RUNI
		9. National Institute of Higher Education, Research, Science & Technology	UNI
Vanuatu	Primary	1. Togale Experimental Station (MOA, Forestry & Fisheries)	GRA
	Secondary	2. Institut de Recherche sur le Café et Cacao	FTRI
		3. Institut de Recherches pour les Huiles et Oleagineux	
Western Samoa	Primary	1. Research and Technical Division, Department of Agriculture	GRA
	Secondary	2. Forestry Division, Department of Agriculture (research unit not identified)	
		3. Institute of Research, Extension & Training in Agriculture	RRO
		4. School of Agriculture, University of South Pacific	RUNI

Source: ISNAR's Small-Countries Data Base.

*_Primary_ refers to those institutes with a primary mandate to conduct research in agriculture. _Secondary_ refers to those institutes whose primary function and activity is not agricultural research but which perform an important function within the NARS.

**Institute types:

GRA   Government agricultural research institutions (national institutes, ministerial departments and divisions, research centers)

GRI   Government scientific research institutes (nonagricultural mandates, institutes of ecology, oceanography, science and technology, etc., working in the food and agricultural sector)

PARA Parastatal research units (commonly attached to commodity boards, associations, and institutes)

NAB   National agribusiness with distinct research departments or units (e.g., Barbados Sugar Industry Ltd. has several research units)

MNAB Standard Brands, Chiquita Brands, Firestone Rubber, Lever Bros. (UNILEVER) have research units based in developing countries

FTRI   French Tropical Research Institute

UNI   University

ARC   Agricultural research council

RRO   Regional agricultural research organization

RUNI Regional university

Annex 3. Number of Researchers in a Sample of 66 Core NARS Institutions in 31 Small Countries

Country	Year	Sci*	PhD	MSc	BSc	Research Institutions/Organization
Barbados	1991	3	3	0	0	West Indies Sugar Cane Breeding Station
		2	0	0	2	Barbados Sugar Cane Variety Testing Station
		3	1	0	2	Sugar Technology Research Unit, Barbados Sugar Industry Ltd.
		2	0	1	1	Agronomy Research Unit, Barbados Sugar Industry Ltd.
		1	1	0	0	Caribbean Agro-Industies Ltd.
Bhutan	1990	19	0	4	15	Research Unit, Dept. of Agriculture
		1	0	0	1	Research & Training Division, Dept. of Forestry
		5	0	3	2	Research Unit, Dept. of Animal Health, MOA
Botswana	1990	36	4	19	13	Department of Agricultural Research, MOA
Burundi	1991	60	2	37	21	Institut des Sciences Agronomiques du Burundi
Cape Verde	1991	20	2	18		Instituto Nacional de Investigação Agraria
Chad	1989	9	3	1	5	Bureau de la Recherche Agronomique
	1991	17	6	11		Laboratoire de Recherches Vétérinaires et Zootechniques
Comoros	1989	7	0	0	7	Research & Development Unit, Federal Centre of Rural Development
Dominica	1991	6	1	2	3	Ministry of Agriculture, Industry, Tourism, Trade, Lands and Surveys
Fiji	1990	17	3	5	9	Research Division of the Dept. of Agriculture
	1989	5	1	4		Research Sect., Animal Health & Production Div.
	1990	6	0	1	5	Sugar Cane Research Institute
Grenada	1991	2	0	0	2	Crop Protection Unit (MoALF&F)
Guyana	1991	28	4	10	14	National Agricultural Research Institute
	1990	3	0	1	2	Research Division, Guyana Sugar Corporation
Honduras	1990	21	0	5	16	Research Dept., Instituto Hondureño del Café
		29	9	9	11	Fundación Hondureña de Investigación Agrícola
		69	2	12	55	Crop Research Department
	1989	48	2	15	31	Centro Nacional de Investigación Forestal Aplicada
Jamaica	1991	27	2	7	18	Dept. of Agricultural Research & Development
		3	1	0	2	Research Department, Jamaican Banana Board
		4	0	2	2	Research Unit, Jamaican Coconut Industry Board
		23	4	8	11	Scientific Research Council
		18	0	4	14	Sugar Research Institute
		3	1	1	1	Research and Ext. Div., Cocoa Industry Board
Laos	1990	19	2		17	Centre National de Recherche Agronomique
Lesotho	1991	25	5	7	13	Agricultural Research Division

Annex 3. (continued)

Country	Year					Institute
Mauritania	1992	3		3		Centre National de l'Elevage et Recherche Vétérinaire
		13		13		Centre National de Recherche Agron. & Develop.
		24		24		Centre National de Recherche Océanographique & des Pêches
Mauritius	1990	52	7	20	25	Mauritius Sugar Industry Research Institute
Namibia	1992	21	0	10	11	Agricultural Research Division, MAWRD
Panama	1991	71	11	37	23	Inst. de Investigación Agropecuaria de Panamá
Papua New Guinea	1992	36	3	13	20	Agric. Research Div., Min. of Agric. & Livestock
		4	1	2	1	University Technology
		4	1	1	2	Wau Ecology Institute
		9	1	2	6	PNG Oil Palm Research Association
		20	3	2	15	PNG Cocoa and Coconut Research Institute
		19	0	6	13	PNG Coffee Research Institute
		3	0	1	2	PNG Sugarcane Research Centre
Paraguay	1991	64	2	22	40	Dept. of Ag., Livestock & Forestry Res. & Ext.
		51		11	40	Instituto Agronomico Nacional
Rwanda	1989	38	1	3	34	Institut des Sciences Agronomiques du Rwanda
São Tomé e Principe	1993	4	0	0	4	Centro de Culturas Alimentares de Mesquita
		3	0	0	3	Agron. Exper. Station, Poto
Sierra Leone	1992	21	4	13	4	Rokupr Rice Research Station
		24	1	13	10	Institute of Agricultural Research, Njala
Somalia	1989	70	0	17	53	National Agricultural Research Institute
St. Lucia	1991	6	0	1	5	Min. of Ag., Lands, Fisheries, For. & Coop.
St. Vincent	1991	1	0	0	1	Research Unit, Department of Agriculture, MOA
Suriname	1989	15	0	9	6	Landbouwproefstation
Swaziland	1990	17	1	9	7	Agricultural Research Division
Togo	1992	2	1		1	Direction Nationale de la Recherche Agron.
	1989	19	2	9	8	Institut National des Cultures Vivières (DRA)
		5	0	1	4	Institut des Plantes à Tubercules
		9	1	4	4	Direction Nutrition et Technologie Alimentaire
		12	1	11	0	Institut National des Sols
		3	2	1	0	Centre de Recherche et d'Elevage d'Avetonou
Trinidad and Tobago	1991	48	3	21	24	Min. of Food Production and Marine Exploitation
	1989	10	1	1	8	Ministry of Environment and National Service
Total		1242	105	464	673	
Average**		18.8	1.6	7.0	10.2	

Note: Data are from 1989 to 1992. Distribution of institutes by year is 13 for 1989; 12 for 1990; 24 for 1991; 17 for 1992. Total=66 institutions in 32 countries.
*Sci refers to number of scientists in the institute.
**Average = number of scientists per institution.

Annex 4. Scope of Research in Small Countries (Topics of Research Regrouped)

Type of Research	Commodity-Based Research Activities					Factor-Based Research		Totals
	GS	TE	MFC	HNE	L &G	SER	NRM	
Crop husbandry or animal husbandry: agronomic investigations (e.g., fertilization, irrigation); seed production; animal feed*	139	30	66	19	24	3	9	290
Plant or animal genetics and breeding: breeding; germplasm collection & evaluation; introduction of new cultivars & breeds	86	22	66	22	22		5	223
Plant or animal protection: pests; diseases; weeds; animal hygiene	40	20	32	14	20	4	11	141
Plant or animal physiology, biochemistry or ecology: growth, development & reproduction	1	7			20			28
Natural resource management: soils, water, forestry, fisheries management**	3	2			1	2	18	26
Other: production economics; postharvest; marketing; storage & transport; product processing & preservation; agricultural machinery; human nutrition	20	5	9	9	19	5	12	79
Totals	289	86	173	64	106	14	55	787
Percent (approx.)	37%	11%	22%	8%	13%	2%	7%	

Notes: Acronyms in subhead refer to the following: GS = Global Staples; TE = Traditional Exports; MFC = Minor Food Crops; HNE = High-Value, Nontraditional Exports; L&G = Livestock and Game; SER = Socioeconomic Research; NRM = Natural Resource Management.
There are possible areas of overlap in the above data with respect to NRM and socioeconomic research. Some of these are explained in the notes below.
*Examples of research activities that would be recorded under NRM or socioeconomics include (a) determining planting density for new forests (NRM) and (b)the economics of planting in rows vs random broadcast.
**NRM research activities could be specific to a particular commodity, for example: (a) soil erosion control in beans by manipulating planting density and time of planting, which would be recorded under beans (GS). The NRM research activities recorded for GS, TE, MFC, HNE, and L are soil related. Those under socioeconomics and NRM are mostly forestry or fisheries research. Similarly, determining stocking rates for a particular type of livestock in a given ecosystem could be recorded under livestock research. (b) The contribution of a legume to soil fertility could be factor based or could address issues of resource management not specific to a particular commodity. An example of this would be soil mapping and classification or assessment of fisheries resources, etc.

BIBLIOGRAPHY

Abe, L.O. and Marcotte, P.L. (1989) *Proceedings of the SACCAR/ISNAR Workshop on Human Resource Management in National Agricultural Research Systems, Harare, Zimbabwe, 2–6 May 1989*. The Hague: International Service for National Agricultural Research.

Adelman, C., Jenkins, D. and Kemmis, S. (1975) *Re-thinking case study: Notes from the second Cambridge conference*. Conferences on Case Study Methods in Educational Research and Evaluation. Cambridge, UK: Churchill College.

Advanced Technology Assessment System. (1992) *Biotechnology and development: Expanding the capacity to produce food*. ATAS Bulletin Issue 9 (Winter 1992). New York: United Nations Department of Economic and Social Development.

Aithnard, T. and Gninofou, A.M. (1992) Analyse de l'évolution du système national de recherche agricole Togolais (1980–1989). The Hague: International Service for National Agricultural Research. *Unpublished ms.*

Anderson, J. (1992) Review of Plucknett, Smith and Ozgediz's book on networking. *Journal of Developing Areas* 26 (2): 268-270.

Antoine, R. and Persley, G. (1992) Strategy for biotechnology research: Mauritius. Paper prepared for the International Workshop on Management Strategies and Policies for Agricultural Research in Small Countries, Réduit, Mauritius, 20 April–2 May 1992. The Hague: International Service for National Agricultural Research.

Argenti, G., Filgueira, C. and Sutz, J. (1990) From standardization to relevance and back again: Science and technology indicators in small, peripheral countries. *World Development* 18(11): 1555-1567.

Arntzen, J. (1993) *Natural resource management and sustainable agriculture: Policy implications for research systems in small developing countries*. ISNAR Discussion Paper 93-02. The Hague: International Service for National Agricultural Research.

Arntzen, J.W., Crowley, E.L. and Eyzaguirre, P.B. (1992) Natural resource management and sustainable agriculture: Implications for NARS in small countries. Paper prepared for the International Workshop on Management Strategies and Policies for Agricultural Research in Small Countries, Réduit, Mauritius, 20 April–2 May 1992. The Hague: International Service for National Agricultural Research.

Baker, R.J. (1992) Agricultural research in Jamaica. Paper prepared for the International Workshop on Management Strategies and Policies for Agricultural Research in Small Countries, Réduit, Mauritius, 20 April–2 May 1992. The Hague: International Service for National Agricultural Research.

Ballantyne, P.G. (1991) *Managing the flow of scientific information for agricultural research in small countries: An issues paper*. ISNAR Small-Countries Study Paper No. 2. The Hague: International Service for National Agricultural Research.

Ballantyne, P.G. (1993) Managing scientific information in agricultural research. *Public Administration and Development* 13, 271-280.

Barampana, D. (1992) Certains aspects de la gestion de la recherche agronomique au Burundi. Paper prepared for the International Workshop on Management Strategies and Policies for Agricultural Research in Small Countries, Réduit, Mauritius, 20 April–2 May 1992. The

Hague: International Service for National Agricultural Research.

Bebbington, A. and Farrington, J. (1993) Governments, NGOs and agricultural development: Perspectives on changing inter-organisational relationships. *Journal of Development Studies* 29(2): 199-219.

Bennell, P. and Oxenham, J. (1983) Skills and qualifications for small island states. *Labour and Society* 8(1): 13-37.

Beye, G. (1992) Strengthening national research systems through regional technical cooperation. Paper prepared for the International Workshop on Management Strategies and Policies for Agricultural Research in Small Countries, Réduit, Mauritius, 20 April–2 May 1992. The Hague: International Service for National Agricultural Research.

Biggs, S.D. (1990) A multiple source of innovation model of agricultural research and technology promotion. *World Development* 18 (11):1481-1499.

Bonte-Friedheim, C.H. (1992) ISNAR and its service to NARS in small countries. Paper prepared for the International Workshop on Management Strategies and Policies for Agricultural Research in Small Countries, Réduit, Mauritius, 20 April–2 May 1992. The Hague: International Service for National Agricultural Research.

Bray, M. (1991) The organization and management of ministries of education in small states. *Public Administration and Development* 11(1): 67-78.

Brown et al. (1990) *State of the world 1990.* New York: W.W. Norton.

Burley, J. (1987) International forestry research networks: Objectives, problems and management. *Unasylva* 39(3/4): 67-73.

Burrill, G.S. and Roberts, W.J. (1992) Biotechnology and economic development: The winning formula. *Bio/technology* 10(6): 647-650.

Byrnes, K.J. (1992) From melon patch to market place: How they learned to export a nontraditional crop. Paper prepared for the International Workshop on Management Strategies and Policies for Agricultural Research in Small Countries, Réduit, Mauritius, 20 April–2 May 1992. The Hague: International Service for National Agricultural Research.

Casas, J. (1992) *Compte-rendu de mission auprès de l'Institut de Recherche Agronomique et Zootechnique (IRAZ) de la CEPGL: Appui à; a préparation du plan quinquennal d'activités.* Rome: Food and Agriculture Organization of the United Nations.

Chong, K. von (1992) Current scope of agricultural research and future strategy in Panama. Paper prepared for the International Workshop on Management Strategies and Policies for Agricultural Research in Small Countries, Réduit, Mauritius, 20 April–2 May 1992 . The Hague: International Service for National Agricultural Research.

Cohen, J.I. (1993) Biotechnology priorities, planning, and policies: A decision-making framework. The Hague: International Service for National Agricultural Research. *Unpublished ms.*

Commandeur, P. (1993) Latin America commences to biotechnologize its industry. *Biotechnology and Development Monitor* 14: 3-5.

Commonwealth Secretariat. (1990) *Basic statistical data on selected countries (with populations of less than 5 million).* London: Economic Affairs Division, Commonwealth Secretariat.

Contreras, M. (1992) *The organization of a small-country agricultural research system with broad research demands: Institutional diversity in Honduras.* ISNAR Small-Countries Study Paper No. 4. The Hague: International Service for National Agricultural Research.

Coolen, P., Beal, G. and Moran, K. (1984) *Case study as a method for field work: Use in selected small island nations.* Honolulu: East-West Center, Institute of Culture and Communication.

Crosson, P. and Anderson, J.R. (1993) *Concerns for sustainability: Integration of natural resource and environmental issues in the research agendas of NARS.* Research Report No. 4. The Hague: International Service for National Agricultural Research.

Crucifix, D. and Packham, J. (1993) *Exporting, risk, and inefficiency.* Tropical Fruits Newsletter No. 6. Port-of-Spain, Trinidad and Tobago: Inter-American Institute for Cooperation on Agriculture.

Daane, J.R.V. and Fanou, J.A. (1986) Faculty building in a small francophone African state: A case study of experiences in Beninese-Dutch inter-university cooperation at the Faculty of Agriculture of the National University of Benin. Paper prepared for the Workshop on Staff Development in Higher Agricultural Education, Malang, Indonesia, 19–23 August 1986. Benin: National University of Benin.

Dahniya, M.T. (1992) An overview of agricultural research in Sierra Leone. Paper prepared for the International Workshop on Management Strategies and Policies for Agricultural Research in Small Countries, Réduit, Mauritius, 20 April–2 May 1992. The Hague: International Service for National Agricultural Research.

Dahniya, M.T. (1993) *Linking science and the farmer: Pillars of the national agricultural research system in Sierra Leone.* ISNAR Small-Countries Study Paper No. 10. The Hague: International Service for National Agricultural Research.

Davies, A. (1991) *Strategic leadership.* London: Woodhead-Faulner Ltd.

Dolman, A. (1985) Paradise lost? The past performance and future prospects of small island developing countries. In *States, microstates, and islands,* edited by E.C. Dommen and P. Hein. Australia: Croom Helm.

Dorji, K., Gapasin, D.P. and Pradhan, P.M. (1992) Integration of research: Bhutan's renewable resource sector. Paper prepared for the International Workshop on Management Strategies and Policies for Agricultural Research in Small Countries, Réduit, Mauritius, 20 April–2 May 1992. The Hague: International Service for National Agricultural Research.

Dregne, M., Kassas, M. and Rozanov, B. (1992) A new assessment of the world status of desertification. *Desertification Bulletin* 1992: 6-18.

Duncan, R.C. (1993) Agricultural export prospects for sub-Saharan Africa. *Development Policy Review* 11 (1): 31-45.

Durant, N. and Blades, H. (1990) An agricultural action plan for the Caribbean in light of the present world market situation. In *Agricultural diversification in the Caribbean.* CARDI/CTA Seminar Proceedings. Ede, The Netherlands: Caribbean Agricultural Research and Development Institute/Technical Centre for Agriculture and Rural Cooperation.

Eaton, N.L. (1991) Midwest Agricultural Biotechnology Information Center: A research and development project. *Agricultural Libraries Information Notes* 17(5): 1-2.

Eicher, C.K. (1988) Food security battles in sub-Saharan Africa. Paper prepared for the VII World Congress for Rural Sociology, Bologna, Italy, 26 June–2 July 1988. The Hague: International Service for National Agricultural Research.

Elliott, H. (1992) Strategies for "workable" NARS in small countries: Establishing a manageable research portfolio. Paper prepared for the International Workshop on Management Strategies and Policies for Agricultural Research in Small Countries, Réduit, Mauritius, 20 April–2 May 1992. The Hague: International Service for National Agricultural Research.

Eyzaguirre, P.B. (1991) *The scale and scope of national agricultural research in small developing countries: Concepts and methodology.* ISNAR Small-Countries Study Paper No. 1. The Hague: International Service for National Agricultural Research.

Eyzaguirre, P.B. (1992) Managing the scale and scope of agricultural research in small countries: An overview. Paper prepared for the International Workshop on Management Strategies and Policies for Agricultural Research in Small Countries, Réduit, Mauritius, 20 April–2 May 1992. The Hague: International Service for National Agricultural Research.

Eyzaguirre, P.B. (1993a) Developing appropriate strategies and organizations for agricultural research in small countries. In *Social science research for agricultural technology development: Spatial and temporal dimensions,* edited by K.A. Dvorak. Wallingford, UK: CAB

International.

Eyzaguirre, P.B. (1993b) The role of agricultural research networks in small countries. *IAALD Quarterly Bulletin* 38(2-3): 69-74.

Eyzaguirre, P.B. and Okello, A.E. (1993) Agricultural research systems in small countries: Implications for public policy and administration. *Public Administration and Development* 13, 233-247.

Eyzaguirre, P.B. and Sivan, P. (1992) Agricultural research in the small countries of the South Pacific: The role of strategic planning in national agricultural research systems. In *Strategic planning for small-country national agricultural research systems (NARS) of the South Pacific islands: Report of a workshop, Apia, Western Samoa, 4–8 March 1991*, edited by A. de S. Liyanage and P. Sivan. Apia, Western Samoa: Institute for Research, Extension and Training in Agriculture.

Fairbairn, T.I.J. (1990) The environment and development planning in small Pacific Island countries. In *Sustainable development and environmental management of small islands*, edited by W. Beller, P. d'Ayala and P. Hein. Paris: United Nations Educational, Scientific and Cultural Organization.

Falconi, C.A. (1993) *Ecuador: Agricultural research in the public and private sectors.* ISNAR Briefing Paper No. 2. The Hague: International Service for National Agricultural Research.

FAO. 1992. Agrostat: Production. [Diskette] Rome: Food and Agriculture Organization of the United Nations.

Faris, D.G. (1991) *Agricultural research networks as development tools: Views of a network coordinator.* Hyderabad, India: International Crops Research Institute for the Semi-Arid Tropics.

Farrington, J. and Bebbington, A. (1991) Institutionalisation of farming systems development—are there lessons from NGO-government links? Paper prepared for the FAO Expert Consultation on the Institutionalisation of Farming Systems Development, Rome, 15–17 October 1991. Rome: Food and Agriculture Organization of the United Nations.

Farrington, J. and Bebbington, A. (1992) From research to innovation: Getting the most from interaction with NGOs in FSR/E. Paper presented at the International Farming Systems Research/Extension Symposium, Michigan State University, 14–18 September 1992.

Farrington, J., Bebbington, A., Lewis, D.J. and Wellard, K. (1993) *Reluctant partners? Non-governmental organizations, the state and sustainable agricultural development.* London: Routledge.

Faye, J. and Bingen, J. (1989) *Sénégal: Organisation et gestion de la recherche sur les systèmes de production à l'Institut Sénégalais de Recherches Agricoles.* OFCOR Case Study No. 6. The Hague: International Service for National Agricultural Research.

Fernandez, F. (1992) The privatization of agricultural research: FHIA's experience in Honduras 1989–1992. Paper prepared for the International Workshop on Management Strategies and Policies for Agricultural Research in Small Countries, Réduit, Mauritius, 20 April–2 May 1992. The Hague: International Service for National Agricultural Research.

Fleming, E. (1995) Research options for high-value exports in South Pacific islands. Discussion Paper No. 95-3. The Hague: International Service for National Agricultural Research..

Forde, B. (1992) The national agricultural research institute model: The case of Guyana. Paper prepared for the International Workshop on Management Strategies and Policies for Agricultural Research in Small Countries, Réduit, Mauritius, 20 April–2 May 1992. The Hague: International Service for National Agricultural Research.

Forsyth, D.J.C. (1990) Technology policy for small developing countries. London: Macmillan.

Foss, P.D. (1990) Building a library network: Simply, affordably, the Pacific way. *Quarterly Bulletin of IAALD* 35 (1):31-34.

Fraley, R. (1992) Sustaining the food supply. *Bio/technology* 10(1): 40-43.

Gakale, L.P. (1992) Current scope of agricultural research in Botswana and the relevance of SACCAR's planning to the national program. Paper prepared for the International Workshop on Management Strategies and Policies for Agricultural Research in Small Countries, Réduit, Mauritius, 20 April–2 May 1992. The Hague: International Service for National Agricultural Research.

George, J.B. (1992) Agricultural research activities in Sierra Leone: A university perspective. Paper prepared for the International Workshop on Management Strategies and Policies for Agricultural Research in Small Countries, Réduit, Mauritius, 20 April–2 May 1992. The Hague: International Service for National Agricultural Research.

Getubig, I.P. Jr., Chopra, V.L. and Swaminathan, M.S. (eds) (1991) *Biotechnology for Asian agriculture: Public policy implications.* Kuala Lumpur, Malaysia: Asian and Pacific Development Centre.

Gilbert, E.H. and Sompo-Ceesay, M.S. (1990) Dealing with the size constraint: Strategies for technology management in small agricultural research systems. In *Methods for diagnosing system constraints and assessing the impact of agricultural research, Vol. I,* edited by R.G. Echeverria. The Hague: International Service for National Agricultural Research.

Gilbert, E.H., Matlon, P. and Eyzaguirre, P.B. (1993) NARS in the small countries of West Africa: New perspectives for vulnerable institutions. The Hague: International Service for National Agricultural Research. *Unpublished ms.*

Goldsworthy, P.R., Eyzaguirre, P.B. and Duiker, S.W. (1995) Collaboration between national, international and advanced research institutes for eco-regional research. In *Eco-regional approaches for sustainable land use and food production,* edited by J. Bouma, A. Kuyvenhoven, B.A.M. Bouman, J.C. Luyten, and H.G. Zandstra. Dordrecht, The Netherlands: Kluwer Academic Publishers.

Graham-Tomasi, T. (1991) Sustainability: Concepts and implications for agricultural research policy. In *Agricultural research policy: International quantitative perspectives,* edited by P.G. Pardey, J. Roseboom and J.R. Anderson. Cambridge, UK: Cambridge University Press.

Guichard, C. (1985) COLEACP—Liaison between the professionals. *The Courier* (92) July-August: 70-73.

Gunn, S. (1993) Last stand for solitary marron. *New Scientist* (1887) 21 Aug: 8.

Hardaker, J.B. and Fleming, E.M. (1989) Agricultural research problems in small developing countries: Case studies from the South Pacific island nations. *Agricultural Economics* 3(4): 279-292.

Hardaker, J.B. and Fleming, E.M. (1990) Agriculture: Key to South Pacific growth. *Partners in Research for Development* (3) April 1990: 26-31.

Hardy, W.F. (1992) Biotechnology laboratory requirements. In *ATAS, biotechnology and development: Expanding the capacity to produce food.* ATAS Bulletin No. 9. New York: United Nations Department of Economic and Social Development.

Hawtin, G. (1991) Foreword. In *Agricultural research networks as development tools: Views of a network coordinator,* by D.G. Faris. Hyderabad, India: International Crops Research Institute for the Semi-Arid Tropics.

Hazelman, S.M. (1992) The Pacific agricultural information system project of the South Pacific Commission. In *Strategic planning for small-country national agricultural research systems (NARS) of the South Pacific islands: Report of a workshop, Apia, Western Samoa, 4–8 March 1991,* edited by A. de S. Liyanage and P. Sivan. Apia, Western Samoa: Institute for Research, Extension and Training in Agriculture.

Hee Houng, M. and Ballantyne, P.G. (1991) *Managing scientific information to meet the changing needs of agricultural research in Trinidad and Tobago.* ISNAR Small-Countries Study Paper No. 3. The Hague: International Service for National Agricultural Research.

Herdt, R.W. (1991) Perspectives on agricultural biotechnology research for small countries.

Journal of Agricultural Economics 42(3): 298-308.

Herdt, R.W. and Lynam, J.K. (1992) Sustainable development and the changing needs of international agricultural research. In *Assessing the impact of international agricultural research for sustainable development: Proceedings of a symposium, Ithaca, NY, 16–19 June 1991*, edited by D.R. Lee, S. Kearl and N. Uphoff. Ithaca, NY: Cornell University Institute for Food, Agriculture and Development.

Hobbs, S.H. (1988) *Product planning tools for agricultural research managers*. ISNAR Staff Note 88-14. The Hague: International Service for National Agricultural Research.

Hobbs, S.H. (1992) Challenges and opportunities for the manager of a small-country national agricultural research system. Paper prepared for the International Workshop on Management Strategies and Policies for Agricultural Research in Small Countries, Réduit, Mauritius, 20 April–2 May 1992. The Hague: International Service for National Agricultural Research.

Hobbs, S.H. and Sachdeva, P.S. (1992) Framework for strategic planning for agricultural research management. In *Strategic planning for small-country national agricultural research systems (NARS) of the South Pacific islands: Report of a workshop, Apia, Western Samoa, 4–8 March 1991*, edited by A. de S. Liyanage and P. Sivan. Apia, Western Samoa: Institute for Research, Extension and Training in Agriculture.

Hodgson, J. (1992) Biotechnology: Feeding the world. *Bio/technology* 10(1): 47-50.

Houssou, M. (1992) Les réseaux et le système national de recherche agricole: Le cas de la Direction de la Recherche Agronomique du Bénin. Paper prepared for the International Workshop on Management Strategies and Policies for Agricultural Research in Small Countries, Réduit, Mauritius, 20 April–2 May 1992. The Hague: International Service for National Agricultural Research.

Howlett, D. (1985) Demographic trends and implications. In *The Pacific in the Year 2000*, edited by R.C. Kiste and R.A. Herr. Honolulu Pacific Islands Studies Program, Centre for Asian and Pacific Studies, University of Hawaii at Manoa, in collaboration with the Pacific Islands Development Program, East-West Center.

IGADD. (1992a) *The IGADD 5-year programme, 1992–1996*. Djibouti: Intergovernmental Authority on Drought and Development.

IGADD. (1992b) *Proceedings of a Workshop on Regional Collaboration in Agricultural Research, Training and Extension in IGADD Countries, Addis Ababa, 2–4 December 1992*. Djibouti: Intergovernmental Authority on Drought and Development.

Ilala, S. (1989) *Coconut industry in Solomon Islands*. APCC Occasional Publication Series No. 13. Jakarta: Asian and Pacific Coconut Community.

INSAH. (1990) *Stratégie et programmation quinquennale 1990–1994*. Bamako, Mali: Institut du Sahel.

INSAH/SPAAR Task Force. (1991) Revitalizing agricultural research in the Sahel: A proposed framework for action. Washington, DC: Special Program for African Agricultural Research. *Unpublished ms.*

Islam, N. (1990) *Horticultural exports of developing countries: Past performances, future prospects, and policy issues*. Research Report No. 80. Washington, DC: International Food Policy Research Institute.

ISNAR. (1989a) *Review of Lesotho's agricultural research system*. ISNAR Country Report No. R48. The Hague: International Service for National Agricultural Research.

ISNAR. (1989b) Guinea-Bissau: An investment in the future. *ISNAR Newsletter* #10. The Hague: International Service for National Agricultural Research.

ISNAR. (1992) *Management of scientific information for agricultural research in small countries: Highlights of a Meeting, Réduit, Mauritius, 20–24 April 1992*. ISNAR Small-Countries Study Paper No. 8. The Hague: International Service for National Agricultural

Research.

ISNAR. (1993) *Agenda 21: Issues for national agricultural research.* ISNAR Briefing Paper No. 4. The Hague: International Service for National Agricultural Research.

Jalan, B. (1982) *Problems and policies in small economies.* London: Croom Helm.

Jallow, A.T. (1992) Regional collaboration in agricultural research: The experience of the Institut du Sahel. Paper prepared for the International Workshop on Management Strategies and Policies for Agricultural Research in Small Countries, Réduit, Mauritius, 20 April–2 May 1992. The Hague: International Service for National Agricultural Research.

Jarret, F.G. and Anderson, K. (1989) *Growth, structural change and economic policy in Papua New Guinea: Implications for agriculture.* National Centre for Development Studies Policy Paper No. 5. Canberra: Australian National University.

Johnson, D.G. (1982) Agricultural research policy in small developing countries. In *Managing renewable natural resources in developing countries,* edited by C.W. Howe. Boulder: Westview.

Jones, K.A. (1990) Classifying biotechnologies. In *Agricultural biotechnology: Opportunities for international development,* edited by G.J. Persley. Wallingford, UK: CAB International.

Kaimowitz, D. (1991) *Cambio tecnológico y la promoción de exportaciones agrícolas no tradicionales en América Central.* San José, Costa Rica: Instituto Interamericano de Cooperación para la Agricultura.

Liyanage, A. de S. and Sivan, P. (eds) (1992) *Strategic planning for small-country national agricultural research systems (NARS) of the South Pacific islands: Report of a workshop, Apia, Western Samoa, 4–8 March 1991.* Apia, Western Samoa: Institute for Research, Extension and Training in Agriculture.

Lynam, J.K. and Herdt, R.W. (1992) Sense and sustainability: Sustainability as an objective in international agricultural research. In *Diversity, farmer knowledge, and sustainability,* edited by J.L. Moock and R.E. Rhoades. Ithaca, NY: Cornell University Press.

Maingot, A.P. (1991) *Small country development and international labor flows: Experiences in the Caribbean.* Boulder: Westview.

Manrakhan, J. (1992) *A century of research: The national agricultural research system of Mauritius.* The Hague: International Service for National Agricultural Research. *Unpublished ms.*

Marcotte, P.L. (1992) Complex research functions: Implications for human resource development and management. Paper prepared for the International Workshop on Management Strategies and Policies for Agricultural Research in Small Countries, Réduit, Mauritius, 20 April–2 May 1992. The Hague: International Service for National Agricultural Research.

Mavuso, M. and Ballantyne, P.G. (1992) *Managing information resources and services for agricultural research in Swaziland.* ISNAR Small-Countries Study Paper No. 9. The Hague: International Service for National Agricultural Research.

Mend, A.F. and Ballantyne, P.G. (1992) *Managing scientific information in a small island nation: The Seychelles experience.* ISNAR Small-Countries Study Paper No. 5. The Hague: International Service for National Agricultural Research.

Meyer, C.A. (1992) A step back as donors shift institution building from the public to the "private" sector. *World Development* 20(8): 1115-1126.

Mills, B. and Gilbert, E. (1989) Agricultural innovation and technology testing by Gambian farmers: Hope for institutionalizing OFR in small country research systems? Paper prepared for the Farming Systems Research/Extension Symposium, Fayetteville, AR, 9–11 October 1989. Banjul, The Gambia: Department of Agricultural Research.

Milner, C. and Westaway, T. (1993) Country size and the medium term growth process: Some cross-country evidence. *World Development* 21(2): 203-211.

Mitchell, G.R. (1992) The changing agenda for research management. *Research Technology*

Management 35(3): 13-21.

Monde, S. S. and Jusu, M.S. (1993) Sustainable plant genetic resources use and management for low-resource rice farming in Sierra Leone. In *The Socio-economic Aspects of Plant Genetic Resources Conservation*. Rome: International Plant Genetic Resources Institute.

Morais, J. (1992) L'information scientifique et technique au Cap Vert. Paper prepared for the International Workshop on Management Strategies and Policies for Agricultural Research in Small Countries, Réduit, Mauritius, 20 April–2 May 1992. The Hague: International Service for National Agricultural Research.

Moustache, A.M. (1992) The role of policy and coordination in orienting the work of donor projects so that they contribute to the research and development needs of the country. Paper prepared for the International Workshop on Management Strategies and Policies for Agricultural Research in Small Countries, Réduit, Mauritius, 20 April–2 May 1992. The Hague: International Service for National Agricultural Research.

Mulongoy, K. (1993) Advanced breeding and plant biotechnology in francophone Africa. *Biotechnology and Development Monitor* (16): 21-22.

Narain, T.M. (1992) Facilitating access to information on agricultural and rural development. Paper prepared for the International Workshop on Management Strategies and Policies for Agricultural Research in Small Countries, Réduit, Mauritius, 20 April–2 May 1992. The Hague: International Service for National Agricultural Research.

Ng Kee Kwong, R. and Ballantyne, P.G. (1992) *Management of scientific information for agricultural research in Mauritius*. ISNAR Small-Countries Study Paper No. 6. The Hague: International Service for National Agricultural Research.

O'Doherty, D. and McDevitt, J. (1991) *Globalisation and the small less advanced member states: Synthesis report*. FAST Occasional Papers No. 291. Brussels: Directorate-General for Science, Research and Development, Commission of the European Communities.

Okello, A.E. and Eyzaguirre, P.B. (1992) *National agricultural research in a regional context: The small countries of Southern Africa*. ISNAR Small-Countries Study Paper No. 7. The Hague: International Service for National Agricultural Research.

Okello, A.E. and T. Namane (1992) *The national agricultural research system of Lesotho*. The Hague: International Service for National Agricultural Research. *Unpublished ms.*

Onanga, M. (1992) Politique scientifique et technologique pour l'agriculture et les ressources naturelles de la République du Congo. Paper prepared for the International Workshop on Management Strategies and Policies for Agricultural Research in Small Countries, Réduit, Mauritius, 20 April–2 May 1992. The Hague: International Service for National Agricultural Research.

Owen, J.M. (1989) Technology, users and the information chain. *International Journal of Information and Library Research* (1)2.

Parasram, S. (1992) National agricultural research systems in the Caribbean: A regional perspective. Paper prepared for the International Workshop on Management Strategies and Policies for Agricultural Research in Small Countries, Réduit, Mauritius, 20 April–2 May 1992. The Hague: International Service for National Agricultural Research.

Pardey, P.G., Roseboom, J. and Anderson, J.R. (1991) *Agricultural research policy: International quantitative perspectives*. Cambridge, UK: Cambridge University Press.

Persaud, B. (1988) Agricultural problems of small states, with special reference to Commonwealth Caribbean countries. *Agricultural Administration and Extension* 29(1): 35-51.

Persley, G.J. (1990) *Beyond Mendel's garden: Biotechnology in the service of world agriculture*. Wallingford, UK: CAB International.

Persley, G.J. (1991) *Agricultural biotechnology: Opportunities for international development*. Wallingford, UK: CAB International.

Persley, G.J. (1992) Biotechnology in agriculture with special reference to developing countries.

In *Strategic planning for small-country national agricultural research systems (NARS) of the South Pacific islands: Report of a workshop, Apia, Western Samoa, 4–8 March 1991,* edited by A. de S. Liyanage and P. Sivan. Apia, Western Samoa: Institute for Research, Extension and Training in Agriculture.

Persley, G.J., Giddings, L.V. and Juma, C. (1993). *Biosafety: The safe application of biotechnology in agriculture and the environment.* Research Report No. 5. The Hague: International Service for National Agricultural Research.

Peshoane, N.T. (1988) Policy on crops production and research in Lesotho. Paper presented at the BLS/SACCAR/ISNAR workshop on Agricultural Research Management, Maseru, Lesotho, 23 May to 3 June 1988.

Peters, T. (1992) *Liberation management: Necessary disorganization for the nanosecond nineties.* London: MacMillan.

Plucknett, D.L., Smith, N.J.H. and Ozgediz, S. (1990) *Networking in international agricultural research.* Ithaca, NY: Cornell University Press.

Pollard, S. (1988) *Atoll economies: Issues and strategy options for development—A review of the literature.* Islands/Australia Working Paper No. 88/5. Canberra: National Centre for Development Studies, Australian National University.

Poon, A. (1990) Flexible specialization and small size: The case of Caribbean tourism. *World Development* (18)1: 109-123.

Reid, J. (1992) Policy definition and research coordination in agriculture: A case for Jamaica. Paper prepared for the International Workshop on Management Strategies and Policies for Agricultural Research in Small Countries, Réduit, Mauritius, 20 April–2 May 1992. The Hague: International Service for National Agricultural Research.

Reid, J. (1993) Analysis of the national agricultural research system in Jamaica: A challenge for policy definition and cooperation. The Hague: International Service for National Agricultural Research. *Unpublished ms.*

Remoortere, F. van and Boer, P. de (1992) Globalization of technology: What it means for American industry. *Research Technology Management* 35(4): 8-9.

Rhoades, R.E. (1984) Understanding small-scale farmers in developing countries: Sociocultural perspectives on agronomic farm trials. *Journal of Agronomic Education* 13: 64-68.

Rive Box, L. de la (1985) Agricultural research policy and organization in small countries: Towards a research agenda. In *Agricultural research policy and organization in small countries.* The Hague: International Service for National Agricultural Research.

Rojas, M. (1993) Agricultural research and nontraditional exports: A case study from Honduras. The Hague: International Service for National Agricultural Research. *Unpublished ms.*

Rudder, W.R. (1992) How policy determines the demand and scope of agricultural research in Trinidad and Tobago. Paper prepared for the International Workshop on Management Strategies and Policies for Agricultural Research in Small Countries, Réduit, Mauritius, 20 April–2 May 1992. The Hague: International Service for National Agricultural Research.

Runge, C.F. (1992) A policy perspective on the sustainability of production environments: Toward a land theory of value. In *Future challenges for national agricultural research: A policy dialogue. Proceedings of the international conference entitled "Challenges and Opportunities for the NARS in the Year 2000: A Policy Dialogue," Berlin, 12–18 January 1992.* The Hague: International Service for National Agricultural Research.

Ruttan, V.W. (1989) Toward a global agricultural research system. In *The changing dynamics of global agriculture,* edited by E.Q. Javier and U. Renborg. The Hague: International Service for National Agricultural Research.

Sabino, A.A. (1992) Instituto Nacional de Investigação Agraria (INIA): Agricultural research in Cape Verde. Paper prepared for the International Workshop on Management Strategies and Policies for Agricultural Research in Small Countries, Réduit, Mauritius, 20 April–2

May 1992. The Hague: International Service for National Agricultural Research.

Santander, V. (1992) Brief prospectus on agricultural research in Paraguay. Paper prepared for the International Workshop on Management Strategies and Policies for Agricultural Research in Small Countries, Réduit, Mauritius, 20 April–2 May 1992. The Hague: International Service for National Agricultural Research.

Sarles, M. (1990) USAID's experiment with the private sector in agricultural research in Latin America and the Caribbean. In *Methods for diagnosing research system constraints and assessing the impact of agricultural research, Vol. 1,* edited by R.G. Echeverria. The Hague: International Service for National Agricultural Research.

SCB. (1983–1991) *Annual reports.* Manzini, Swaziland: Swaziland Cotton Board.

Schahczenski, J.J. (1990) Development administration in the small developing state: A review. *Public Administration and Development* 10 (1): 69-80.

Schuh, G.E. and G.W. Norton (1992) Agricultural research in an international policy context. In *Agricultural research policy: International quantitative perspectives,* edited by P.G. Pardey, J. Roseboom and J.R. Anderson. Cambridge, UK: Cambridge University Press.

Schwarz, C. (1992) Brief notes on the national agricultural research system of Guinea-Bissau. Paper prepared for the International Workshop on Management Strategies and Policies for Agricultural Research in Small Countries, Réduit, Mauritius, 20 April–2 May 1992. The Hague: International Service for National Agricultural Research.

Singh, R.B. (1992) Regional cooperation in agricultural research in the Pacific island countries. In *Strategic planning for small-country national agricultural research systems (NARS) of the South Pacific islands: Report of a workshop, Apia, Western Samoa, 4–8 March 1991,* edited by A. de S. Liyanage and P. Sivan. Apia, Western Samoa: Institute for Research, Extension and Training in Agriculture.

Sison, J.C. (1990) The implementation and management of networks. *Quarterly Bulletin of IAALD* 35 (4): 187-195.

Sitapai, E.C. (1992) The national agricultural research system in Papua New Guinea. Paper prepared for the International Workshop on Management Strategies and Policies for Agricultural Research in Small Countries, Réduit, Mauritius, 20 April–2 May 1992. The Hague: International Service for National Agricultural Research.

Sivan, P. (1992a) Agricultural research in Fiji. Paper prepared for the International Workshop on Management Strategies and Policies for Agricultural Research in Small Countries, Réduit, Mauritius, 20 April–2 May 1992. The Hague: International Service for National Agricultural Research.

Sivan, P. (1992b) The future of agricultural research in the South Pacific. In *Strategic planning for small-country national agricultural research systems (NARS) of the South Pacific islands: Report of a workshop, Apia, Western Samoa, 4–8 March 1991,* edited by A. de S. Liyanage and P. Sivan. Apia, Western Samoa: Institute for Research, Extension and Training in Agriculture.

Sivan, P. and Eyzaguirre, P.B. (1991) Agricultural research in the Pacific. *Courier* (130): 94-96.

Small, W. (1992) The organizations contributing to agricultural research and development in Barbados. Paper prepared for the International Workshop on Management Strategies and Policies for Agricultural Research in Small Countries, Réduit, Mauritius, 20 April–2 May 1992. The Hague: International Service for National Agricultural Research.

Stonich, S.C. (1991) The promotion of non-traditional agricultural exports in Honduras: Issues of equity, environment and natural resource management. *Development and Change* 22: 725-755.

Suh, J.B., Eckebil, J.P., Deganus, E.F., Uriyo, A.P., Whyte, J.B.A. and Ogunyinka, O.M. (1992) Mechanisms for collaborating and strengthening NARS in sub-Saharan Africa: The research liaison scientist scheme. Paper prepared for the International Workshop on Management

Strategies and Policies for Agricultural Research in Small Countries, Réduit, Mauritius, 20 April–2 May 1992. The Hague: International Service for National Agricultural Research.

Tal, E. (1985) R&D in Panama: R&D policy in a small developing country. *Science and Public Policy* 12(5): 253-263.

Taylor, T.A. (1992) Organizational options for small national agricultural research systems. Paper prepared for the International Workshop on Management Strategies and Policies for Agricultural Research in Small Countries, Réduit, Mauritius, 20 April–2 May 1992. The Hague: International Service for National Agricultural Research.

Thottappilly, G., Monti, L., Mohan Raj, D.R. and Moore, A.W. (eds) (1992) *Biotechnology: Enhancing research on tropical crops in Africa.* Ibadan, Nigeria: International Institute of Tropical Agriculture.

Trigo, E.J. (1987) Agricultural research in small countries: Some organizational alternatives. Paper prepared for the NIHERST Agricultural Research Seminar, Centeno, Trinidad, 1–3 October 1987. San José, Costa Rica: Instituto Interamericano de Cooperación para la Agricultura.

Turagakula, E. (1992) Current scope of the research division within the Ministry of Primary Industries, Republic of Fiji. Paper prepared for the International Workshop on Management Strategies and Policies for Agricultural Research in Small Countries, Réduit, Mauritius, 20 April–2 May 1992. The Hague: International Service for National Agricultural Research.

UNCTAD. 1992. UNCTAD commodity yearbook 1992. New York/Geneva: United Nations Conference on Trade and Development.

Valverde, C. (1988) *Agricultural research networking: Development and evaluation.* ISNAR Staff Note No. 88-2b. The Hague: International Service for National Agricultural Research.

Versteeg, M. (1992) The technology-transfer unit at the Benin station of IITA: Support to the Benin national agricultural research system. Paper prepared for the International Workshop on Management Strategies and Policies for Agricultural Research in Small Countries, Réduit, Mauritius, 20 April–2 May 1992. The Hague: International Service for National Agricultural Research.

Wagner, C.K. (1992) International R&D is the rule. *Bio/technology* 10(5): 529-531.

Walmsley, D. (1990) *Proceedings of Agricultural Diversification in the Caribbean, Barbados, 27 November–1 December 1989.* St. Augustine, Trinidad and Tobago: Caribbean Agricultural Research and Development Institute.

Walsh, V. (1987) Technology, competitiveness and the special problems of small countries. *STI Review* 2: 81-133.

WARDA. (1988) A decade of mangrove swamp rice research. Bouaké, Côte d'Ivoire: West African Rice Development Association.

WCED. (1987) *Our common future.* Oxford: Oxford University Press for World Commission on Environment and Development.

Wendt, F.S. (1986) Organisational and structural effectiveness in the bureaucracies of agriculture in small island countries: Can government routine effect agricultural development? *Alafua Agricultural Bulletin* 11(2): 1-15.

Williams, E.T. (1992) The formulation of agricultural research policy in Dominica: Demand and scope. Paper prepared for the International Workshop on Management Strategies and Policies for Agricultural Research in Small Countries, Réduit, Mauritius, 20 April–2 May 1992. The Hague: International Service for National Agricultural Research.

Wilson, G.F. (1992) Agricultural research in the private sector: The agricultural research program of the Jamaica Agricultural Development Foundation. Paper prepared for the International Workshop on Management Strategies and Policies for Agricultural Research in Small Countries, Réduit, Mauritius, 20 April–2 May 1992. The Hague: International Service for National Agricultural Research.

Wilson, L.A. and Singh, R. (1987) The planning and management of agricultural research at the University of the West Indies. In *The planning and management of agricultural research in the South Pacific, Report of a Workshop*. The Hague: International Service for National Agricultural Research.

World Bank. (1990) *Kingdom of Tonga, agricultural sector strategy review*. Washington, DC: The World Bank.

World Bank. (1991) *Western Samoa, agricultural sector strategy review*. Washington, DC: The World Bank.

World Bank. (1995) *World Tables 1995* [Diskette]. Washington, DC: World Bank.

Wyeth, J. (1989) *Diversification: Eight lessons from Honduran experience in the coffee sector*. IDS Discussion Paper No. 259. Brighton, UK: Institute of Development Studies.

Yin, R.K. (1984) *Case study research: Design and methods*. Beverly Hills: Sage Publications.

Zacarías, C. (1992) Agricultural diversification policies, high-value nontraditional exports, and the role of socioeconomic and marketing research in Honduras. Paper prepared for the International Workshop on Management Strategies and Policies for Agricultural Research in Small Countries, Réduit, Mauritius, 20 April–2 May 1992. The Hague: International Service for National Agricultural Research.

Acronyms and Abbreviations

AARINENA	Association of Agricultural Research Institutions in the Near East and North Africa
ACIAR	Australian Centre for International Agricultural Research
ACP	Africa, Caribbean, and Pacific/European Union—Lomé Convention
ACU	Association of Commonwealth Universities
AgGDP	agricultural gross domestic product
AIDAB	Australian International Development Assistance Bureau
APAARI	Asian and Pacific Asssociation of Agricultural Research Institutions
ARD	Agricultural Research Division—Lesotho
ASEAN	Association of South-East Asian Nations
AVRDC	Asian Vegetable Research and Development Centre
CACM	Central American Common Market
CaIB	Cocoa Industry Board—Jamaica
CARDI	Caribbean Agricultural Research and Development Institute
CARICOM	Caribbean Community and Common Market
CARIS	Current Agricultural Research Information System
CATIE	Centro de Agricultura Tropical de Investigación y Enseñanza
CCTA	Commission de Coopération Technique en Afrique au Sud du Sahara
CEAO	Communauté Economique Afrique Ouest
CEDC	Caribbean Export Development Corporation
CEEMAT	Centre d'Etudes et d'Expérimentation du Machinisme Agricole Tropical—France
CENIFA	Centro Nacional de Investigación Forestal Aplicada
CEPGL	Communauté Economique des Pays des Grands Lacs
CFC	Caribbean Food Corporation
CGA	Citrus Growers Association—Jamaica
CGIAR	Consultative Group on International Agricultural Research
CGRRT	Regional Coordination Centre for Research and Development of Coarse Grains, Pulses, Roots and Tuber Crops
CIB	Coffee Industry Board—Jamaica
CILSS	Comité Inter-Etats de la Lutte Contre la Sécheresse au Sahel
CIMMYT	Centro Internacional de Mejoramiento de Maíz y Trigo
CIP	Centro Internacional de la Papa
CIRAD	Centre de Coopération Internationale en Recherche Agronomique pour le Développement
CNES	Centre National d'Etudes des Sols—Congo
CNP	Centre National de Pisìculture—Congo
CNRST	Conseil National de la Recherche Scientifique et Technique
CoIB	Coconut Industry Board—Jamaica

COLEACP	Comité de liaison des fruits tropicaux et légumes de contre saison originaires des états ACP
CORAF	Conference des Responsables de la Recherche Agronomique Africains
CREAT	Centre de Recherche et d'Elevage d'Avétonou—Togo
CSA	Conseil Scientifique pour l'Afrique Sud du Sahara
CTA	Technical Centre for Agricultural and Rural Cooperation
CTFT	Centre Technique Forestier Tropical—CIRAD
DEPA	Departamento de Estudos e Pesquisa Agrícola—Guinea-Bissau
DGDR	Direction Générale du Développement Rural
DICTA	Dirección de Ciencia y Tecnología Agropecuaria—Honduras
DIP	Departamento de Investigación Pecuaria
DIPESCA	Dirección General de Pesca y Acuacultura
DNRA	Direction Nationale de la Recherche Agronomique—Togo
DNTA	Direction de la Nutrition et de la Technologie Alimentaire—Togo
DPV	Direction de la Protection des Végétaux—Togo
DSA	Departement Systèmes Agraires
EAP	Escuela Agrícola Panamericana
ECOWAS	Economic Community of West African States
EC	European Community
ESA	Ecole Supérieure d'Agronomie—Togo
ESCAP	Economic and Social Commission for Asia and the Pacific
ESNACIFOR	Escuela Nacional de Ciencias Forestales
FAO	Food and Agriculture Organization of the United Nations
FARC	Food and Agriculture Research Council—Mauritius
FHIA	Fundación Hondureña de Investigación Agrícola
FUSADES	Fundación Salvadoreña para el Desarrollo
GERDAT	Groupement Etudes Recherches Développement Agronomie Tropicale
GRP	Green Revolution Program—Sierra Leone
GTZ	Gesellschaft für Technische Zusammenarbeit—Germany
IADP	integrated agricultural development project
IAR	Institute of Agricultural Research—Sierre Leone
IARC	international agricultural research center
IBS	Intermediary Biotechnology Service—ISNAR
ICLARM	International Centre for Living Aquatic Resources Management
ICRISAT	International Crops Research Institute for the Semi-Arid Tropics
IDIAP	Instituto de Investigación Agropecuaria de Panamá
IDRC	International Development Research Centre—Canada
IEMVT	Institut d'Elevage et de Médecine Vétérinaire des Pays Tropicaux—CIRAD
IER	Institut d'Economie Rurale
IGADD	Intergovernmental Authority on Drought and Development
IHCAFE	Instituto Hondureño del Café
IICA	Instituto Interamericano para la Cooperación en la Agricultura
IITA	International Institute of Tropical Agriculture
ILCA	International Livestock Centre for Africa
IMBO	Institute of Marine Biology and Oceanography
INEAC	Institut National pour l'Etude Agronomique du Congo
INIBAP	International Network for the Improvement of Banana and Plantain
INIP	Instituto Nacional de Investigação Pescária

INPA	Instituto Nacional de Pesquisa Agricola
INPT	Institut des Plantes à Tubercules—Togo
INS	Institut National des Sols—Togo
INSAH	Institut du Sahel
IPR	intellectual property rights
IRAT	Institut de Recherches Agronomiques Tropicales et des Cultures Vivrières—CIRAD
IRAZ	Institut de Recherche Agronomique et Zootechnique—Pays des Grands Lacs
IRCA	Institut de Recherches sur le Caoutchouc—CIRAD
IRCC	Institut de Recherche du Café et du Cacao—Togo
IRCC	Institut de Recherche du Café, du Cacao et Autres Plantes Stimulantes—CIRAD
IRCT	Institut de Recherche du Coton et des Textiles Exotique—CIRAD
IRETA	Institute of Research, Extension, and Training in Agriculture—University of South Pacific
IRFA	Institut de Recherches sur les Fruits et Agrumes—CIRAD
IRHO	Institut de Recherche pour les Huiles et Oléagineux—CIRAD
ISABU	Institut des Sciences Agronomiques du Burundi
ISNAR	International Service for National Agricultural Research
ISRA	Institut Sénégalais de Recherches Agricoles
ISSCT	International Society of Sugar Cane Technologists
ITC	International Trypanotolerance Centre
IUCN	International Union for the Conservation of Nature and Natural Resources
JADF	Jamaica Agricultural Development Foundation
JARP	Jamaica Agricultural Research Project
JICU	Junta de Investigaçãoes Científica do Ultramar
LNS	Laboratoire National d'Analyse des Sols—Mauritania
LWP	Lesotho Woodlot Project
MAFNR	Ministry of Agriculture, Fisheries and Natural Resources—Mauritius
MANRF	Ministry of Agriculture, Natural Resources and Forestry—Sierra Leone
MERCOSUR	Mercado Comun del Sur
MPI	Ministry of Primary Industries—Fiji
MSIRI	Mauritius Sugar Industry Research Institute
NARCC	National Agricultural Research Coordinating Council—Sierra Leone
NARI	National Agricultural Research Institute—Guyana
NARS	national agricultural research system
NGO	nongovernmental organization
NIR	National Institute of Development Research and Documentation—Botswana
NRM	natural resource management
OAS	Organization of American States
OAU	Organization of African Unity
ODI	Overseas Development Institute
OFR	on-farm research
ORSTOM	Office de la Recherche Scientifique et Technique d'Outre-Mer—France
PROCACAO	Programa Cooperativo de Investigación y Transferencia de Tecnología Agropecuaria para el Cacao
PROCIANDINO	Programa Cooperativo de Investigación Agrícola para la Subregión Andina

PROCICARIBE	Programa Cooperativo de Investigación y Transferencia de Tecnología Agropecuaria para la Subregión de Caribe
PROCICENTRAL	Programa Cooperativo de Investigación y Transferencia de Tecnología Agropecuaria para la Subregión de Centro América
PROCISUR	Programa Cooperativo de Investigación Agrícola para la Subregión del Cono Sur
RESADOC	Regional Documentation Network—Institut du Sahel
RRS	Rice Research Station—Sierra Leone
RSTCA	Regional Sugarcane Training Centre for Africa
SACCAR	Southern African Centre for Cooperation in Agricultural Research
SADC	Southern African Development Community
SADCC	Southern Africa Development Co-ordination Conference
SAFGRAD	Semi-Arid Food Grain Research and Development
SCB	Swaziland Cotton Board
SCMA	Standing Committee of Ministers responsible for Agriculture
SIRI	Sugar Industry Research Institute—Jamaica
SPAAR	Special Program for African Agricultural Research
SPC	South Pacific Commission
SPF	South Pacific Forum
SPID	Storage and Prevention Infestation Division—Jamaica
SRC	Sugar Cane Research Centre—Fiji
	Scientific Research Council—Jamaica
UDEAC	Union Douanière et Economique de l'Afrique Centrale
UNCTAD	United Nations Conference on Trade and Development
UNDP	United Nations Development Programme
UNEP	United Nations Environment Programme
USAID	United States Agency for International Development
USP	University of the South Pacific
UWI	University of the West Indies
WARDA	West African Rice Development Association
WCED	World Commission on Environment and Development
WICBS	West Indies Sugar Cane Breeding Station
WINBAN	Windward Islands Banana Growers' Association

INDEX